# MEETING *the* FAMILY

*For James and Anna,*

*two of the most amazing people I know*

# MEETING *the* FAMILY

## ONE MAN'S JOURNEY THROUGH HIS HUMAN ANCESTRY

**DONOVAN WEBSTER**
*with a foreword by* **SPENCER WELLS**

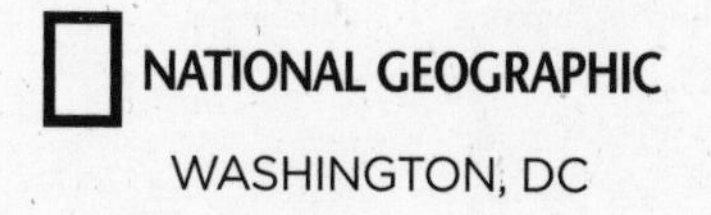


WASHINGTON, DC

Published by the National Geographic Society
1145 17th Street N.W., Washington, D.C. 20036


Library of Congress Cataloging-in-Publication Data
Webster, Donovan.
Meeting the family : one man's journey through his human ancestry / Donovan Webster ; with a foreword by Spencer Wells.
p. cm.
Includes bibliographical references.
ISBN: 978-1-4262-0573-6 (hardcover) -- ISBN 978-1-4262-0604-7 (e-book)
1. Human population genetics. 2. Human beings--Migrations. 3. Genealogy. 4. Heredity.
5. Webster, Donovan. I. Title.
GN289.W43 2010
929'.20973--dc22
2009050471

Photo Credits:
Steve McCurry: Cover, 12, 26, 110, 165, 182, 192, 242. Shutterstock: 74, 82.
National Geographic Stock/Martin Gray: 120. iStockphoto/Klaas Lingbeek-van Kranen: 296.

The National Geographic Society is one of the world's largest nonprofit scientific and educational organizations. Founded in 1888 to "increase and diffuse geographic knowledge," the Society works to inspire people to care about the planet. It reaches more than 325 million people worldwide each month through its official journal, *National Geographic,* and other magazines; National Geographic Channel; television documentaries; music; radio; films; books; DVDs; maps; exhibitions; school publishing programs; interactive media; and merchandise. National Geographic has funded more than 9,000 scientific research, conservation and exploration projects and supports an education program combating geographic illiteracy.

For more information, please call 1-800-NGS LINE (647-5463) or write to the following address:

National Geographic Society
1145 17th Street N.W.
Washington, D.C. 20036-4688 U.S.A.

Visit us online at www.nationalgeographic.com

For information about special discounts for bulk purchases, please contact
National Geographic Books Special Sales: ngspecsales@ngs.org

For rights or permissions inquiries, please contact National Geographic Books
Subsidiary Rights: ngbookrights@ngs.org

*Interior design: Cameron Zotter*

Printed in U.S.A.

10/WCPF-CML/1

# CONTENTS

# MEETING *the* FAMILY

## ONE MAN'S JOURNEY THROUGH HIS HUMAN ANCESTRY

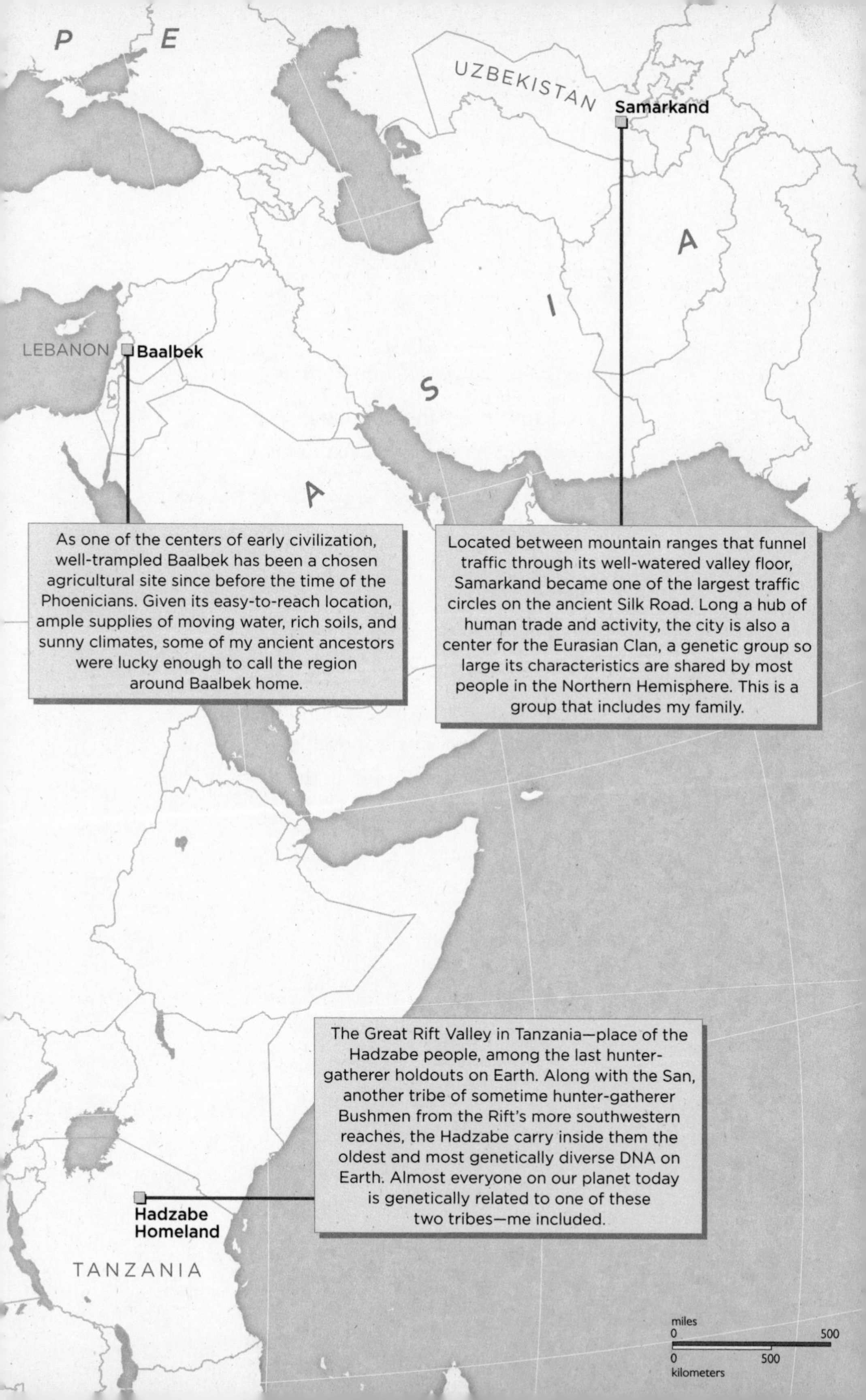

P
E
UZBEKISTAN
Samarkand
A
I
S
A
LEBANON
Baalbek
As one of the centers of early civilization, well-trampled Baalbek has been a chosen agricultural site since before the time of the Phoenicians. Given its easy-to-reach location, ample supplies of moving water, rich soils, and sunny climates, some of my ancient ancestors were lucky enough to call the region around Baalbek home.
Located between mountain ranges that funnel traffic through its well-watered valley floor, Samarkand became one of the largest traffic circles on the ancient Silk Road. Long a hub of human trade and activity, the city is also a center for the Eurasian Clan, a genetic group so large its characteristics are shared by most people in the Northern Hemisphere. This is a group that includes my family.
The Great Rift Valley in Tanzania—place of the Hadzabe people, among the last hunter-gatherer holdouts on Earth. Along with the San, another tribe of sometime hunter-gatherer Bushmen from the Rift's more southwestern reaches, the Hadzabe carry inside them the oldest and most genetically diverse DNA on Earth. Almost everyone on our planet today is genetically related to one of these two tribes—me included.
Hadzabe Homeland
TANZANIA
miles
0
500
0
500
kilometers

*Homo sum, humani nil a me alienum puto.*

I am a man, and nothing that
concerns humanity is alien to me.

*Terence, Roman dramatist, second century* B.C.

*Si más personas llegan, son nada.*
*Hasta que se imposible, echamos más agua en la sopa.*

If more people arrive, it's nothing.
Until it becomes impossible,
we'll just throw more water in the soup.

*Old Mexican saying*

# FOREWORD

BY SPENCER WELLS, *Director of the Genographic Project*

**BLACK, BROWN, AND WHITE;** short and tall; blond, redhead, and brunette. We are an incredibly diverse species, almost dizzying in the variety of surface features that define us. Traveling the world, or simply walking down the street in a large city, we encounter people who seem to be so different from ourselves and from one another. But just how different are we? Appearances, it turns out, are deceiving. Peer beneath the surface, down inside our cells, into our DNA, and you'll get a very different story. It turns out that humans are nearly identical at the genetic level. Yes, that's right, your DNA is basically the same as everyone else's on the planet, and that makes you part of a much larger family than you ever suspected.

In this remarkable book, Don Webster traces his ancestral journey, following the path taken by his DNA from an African homeland around 60,000 years ago, into the Middle East, Central Asia, and ultimately to his English ancestral homeland.

Along the way he meets some pretty remarkable people, from Julius the Hadzabe chief to his distant relations in Uzbekistan. It's a travelogue with a twist, one in which the concept of family is reevaluated as Don expands the scope of his own family tree.

As I write this, the National Geographic Society's Genographic Project is in its fourth year. Launched in 2005, using the tools of molecular genetics, its goal has been to track the ancient migrations of our ancestors, using modern science to gain insight into some of humanity's oldest questions. It was a scientific quest, a global effort to disentangle the stories that weave us together as members of the human family.

Scientific inquiry was always what motivated us to create the project, but the public at large has thronged to the project as well. Over 310,000 people have swabbed their cheeks and sent the samples off to our laboratories, adding their own chapters to the human history. This response has been extraordinary—we had expected to attract perhaps 100,000 public participants during the course of the project—and it gives us extraordinary power to decipher the scientific details of our species' past. But it also, inadvertently, has created a community, albeit one spread across more than 130 countries on every inhabited continent.

Donovan Webster's goal was to meet some of the members of that community. His project began with an article commissioned for *National Geographic Traveler* magazine in 2005. Don was one of the very first people to be tested by the Genographic Project, and he chose to look at his Y chromosome. Tracing a purely paternal line, the pattern of markers in Donovan's DNA told a story of an African beginning, hunting and gathering on the savannas of the Rift Valley in East Africa. Donovan's ancestors left Africa around 50,000 years ago as part of the second wave of migration into Eurasia, via the Middle East. From there they zigzagged across the steppes of Central Asia before entering Europe around 35,000 years ago. This grand journey was what Don was hoping to retrace when we met in my office at National Geographic headquarters in Washington, D.C., just before the launch of the Genographic Project. I plotted the key locations on a map for him, from the remote wilds of Tanzania to the Basque country in northern Spain—places his own ancestors had traversed tens of thousands of years ago. This skeletal outline was what he was hoping to put flesh on with his own travels, meeting his distant cousins along the way.

Don's story is at once unique and universal. It intersects with everyone else's at various points along the way. Each of us is carrying the pattern of our family's wanderings inside ourselves, in our own DNA. The tools of modern molecular genetics can help us read the history encoded there. Written in the pattern of nucleotides, the As, Cs, Gs, and Ts in our genome tell a story of birth, death, adversity, and triumph—an epic trek from an African homeland to the far corners of the Earth. The baroque tapestry of human diversity, woven from such seemingly disparate threads, ultimately reveals a deeper pattern of shared journeys. We are truly members of a human family far more closely related than we ever dreamed of only a generation ago—one that includes Europeans, Asians, Native Americans, New Guineans, Africans, and everyone in between.

So read on—let's meet the family.

*A young Hadzabe tribesman, one of my ancient relatives, in the Great Rift Valley of Tanzania*

# PROLOGUE

## THE MIRACLE AT 36,000 FEET

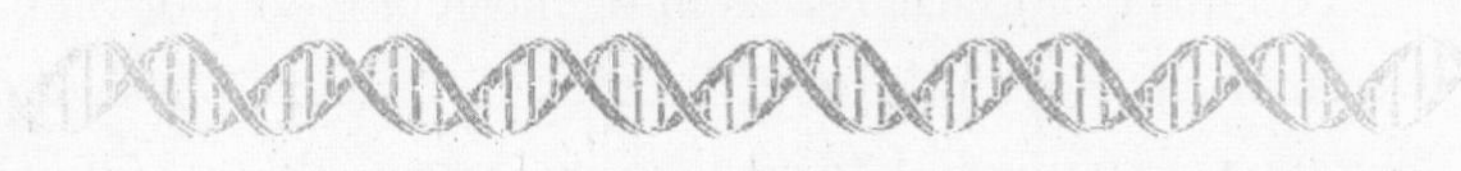

**NO MATTER WHO YOU ARE** and no matter where you come from, the Three Big Questions eventually bubble up.

At the moment, as I get ready to tackle those questions once again, I'm belted into an aisle seat inside a United Airlines jet, halfway across the Atlantic Ocean: on a GPS-determined track between Washington, D.C., and Frankfurt, Germany. A minute ago, in advance of our first in-flight meal, an attendant placed a glass of tasty French Bordeaux on the armrest of my seat. For dinner, she has informed me, I can choose between steak with garlic mashed potatoes and salad, chicken with a Florentine-style pasta and a salad, or a Japanese-style meal of soup, nori rolls, and tempura shrimp and vegetables over steamed rice.

A moment after the wine arrived, and with the press of a button, a video flat-screen has emerged from the royal-blue leather arm of my chair. The screen announces that our selection of in-flight movies will begin in three minutes.

In the meantime, I'm listening to my iPod, a black-anodized box a quarter the size of my hand that contains a survey of Western music: from ars antiqua and Gregorian chant to Jack's Mannequin and last month's new Oasis song. The offerings in this little box make protracted stops in the neighborhoods of Monteverdi, Bach, Haydn, Mozart, Beethoven, Brahms, Gershwin, Ellington, and Copeland. In my iPod are also outlying suburbs of Hank Williams, Lennon/McCartney, Thomas Mapfumo, James Taylor, Marvin Gaye, Ani DiFranco, Miles Davis, and Beck. In all, at this moment, my iPod contains 3,626 pieces of music . . . and it could hold more. Yet outwardly it's just a wafer of matte-black metal that weighs almost nothing and hooks its musical trove into my ears through a slim white headphone cord.

How is this possible? That's a question for later. I mean, we could ponder it more right now, but the movie's about to start.

Greetings from 2010: a world where, for a large and still growing percentage of people, we can get almost anything we want . . . instantaneously. Even in a pressurized jet cabin at 36,000 feet.

Right now, nearly seven miles above the Earth's surface, an array of hot food and cold drinks—plus the history of music and a broad selection of popular and classic movies—is at our disposal, available.

Want a glass of icy, potable water at 36,000 feet? Not a problem.

Want a cup of hot, fresh-brewed coffee? Sure. An apple or a bunch of red grapes . . . plus a quarter round of Camembert presented on a small, white china plate? No worries. These very items are carried on a stainless-steel rolling cart draped in crisp white linen that—in a few minutes—will trundle up the aisle.

But here's the flip side of all this immediacy and ease. Just beneath its surface, it's both mind-spinningly sophisticated and far more expensive than is outwardly comprehensible. It isn't simply dozens of luxury desires being addressed simultaneously in a pressurized aircraft jetting across the world's second largest ocean. It's a whole world of complexity.

The icy bottled water was carried aboard in clear plastic containers imported from France. The coffee? Grown in Colombia. The apple could well be off a tree in New Zealand; then, after picking, it was trucked to the national capital at Christchurch, shipped by refrigerated cargo container to Washington, D.C., . . . where it was loaded onto this specific flight. The grapes? Given the season, they're most likely from Peru. The Camembert? Don't know exactly, . . . but we could ask. Still, it's notable that, because of persnickety French gustatory law, Camembert used to be made only in northwest France, in a few districts inside Normandy to be precise. But these days, once away from the constraints of French laws, this particular "Camembert style" cheese could come from about anywhere.

And yet the total contents of this high-tech aircraft somehow came to be amassed miraculously in one spot—on the tarmac at Washington Dulles International Airport—to be larded aboard this 767 for this specific flight.

Still, what's most amazing, the root source of this Miracle at 36,000 Feet, isn't that all this is merely possible or that it's happening simultaneously aboard thousands of similar aircraft in transit around the world. Instead, it's this: my great-grandparents (and their great-grandparents before them) could never have imagined such a scenario. Even my parents, now in their 70s, arrived in a world where things like touch-screen video weren't realistically conceivable; where having a laptop computer and an MP3 player at their disposal was possible only in fiction or "Dick Tracy" comics.

Hell, as kids, people of my now-middle-aged generation couldn't have imagined it, either.

But now, it's all here, having risen up in only the last two decades: less time than it takes an oak tree to mature.

In fact, until the last decade of the 20th century, even the notion of cold springwater from France recrossing the Atlantic on a jet made in Seattle accompanied by a private Mozart soundtrack digitally recorded in Cleveland and loaded onto a California-designed device manufactured in China . . . well, all of it sat w-aaaaay beyond the realm of realistic consideration.

And this is before we get to the comfy leather seat, the climate-controlled cabin, and the smoothly warm flavor of a nice Bordeaux that's already crossed the Atlantic once from France to arrive in Washington, D.C., only to be put back aboard a jet, returning to Europe.

In the last few decades, the world has grown smaller than at any other time in human history. And people and events on this planet are now moving across it with a speed and on a scale that hasn't occurred at any other time in history.

---

# PROLOGUE

It was February 1976, and as part of an American Bicentennial project put on by the U.S. Congress, I was a high school sophomore fortunate enough to interview a gentleman named Gordon Anderson: a man who, that same year, had turned 100.

Mr. Anderson had grown up on the plains of Kansas, and he could remember the isolated clumps and remnant herds of bison that lived on the prairies and plains beyond his small town. When he and his family traveled in the 1880s and 1890s, they did so by horse or, with heavier goods to transport, horse wagon or oxcart. When we met, it was in a hospital in Chicago, where he was hooked to an electronic heart monitor and had a clear plastic tube leading from an intravenous-drip bottle to a needle inserted into a vein on the back of his hand.

"I can't give you much in the way of wisdom," he said. "But I can tell you this. The most amazing thing I ever saw was as a young man: it was the first electric streetlight in my town. A naked glass bulb, hanging on a length of wire from a wooden pole in the night, lighting the darkness. That we had been able to do this on the plains of Kansas was the single most amazing thing I've seen in 100 years. Well, that and the walk Neil Armstrong took on the moon in 1969, when he made those first steps on live TV. We put a man on the moon . . . and I saw it happening on television."

Inside this place of pale fluorescent light, sterile white paint, and institutional bedding, his room in Chicago's Wesley Memorial Hospital, Mr. Anderson paused for a long moment. It was hard to know if he was going to smile—or cry.

"And to think," he finally said, "both of these things happened in a single life. My lifetime. Well, that's just incredible."

Mr. Anderson had a point.

This book is about the rise of modern humans and how they've come to dominate the Earth.

In it, we'll look at how *Homo sapiens,* as a species, has progressed over millions of years; how we've sometimes fallen down—spectacularly on a few occasions—and how all of this history exists in one strand of DNA (deoxyribonucleic acid), the ultimate intergenerational gift.

This substance called DNA lives inside each of our cells. It is made up of 46 chromosomes—with 23 having been passed along from your mother, and 23 from your father—and these chromosomes are your body's blueprint, its instruction manual, and for 99.9 percent of everyone on Earth, the instructions contained in the genes they carry are exactly the same. What this means is that, even though the sample of DNA examined for this book was taken from me—by scraping some cells off the inside of my cheek on a cold and gray February morning—its source is almost immaterial. An overwhelming amount of the DNA in my individual human sample is the same DNA that's in all of us. And for each of us, this set of chromosomes containing directions for growth and the creation of proteins and enzymes and brain function and susceptibility to certain diseases has been handed along through our ancestors' gift of their own DNA—a heritage from a long time ago.

Of course, even a 99.9 percent similarity still leaves a tiny 0.1 percent remainder of our DNA that is individual. And inside this tiny genetic shard is a whole world of differentiation: the basis not only for physical variations between, say, Han Chinese and Namibians, but also a history of how each of our own ancestors migrated across and around the planet. And, believe it or not, there's a clock in there, too; which pretty much defines when our ancestors were in these varied places.

My sample of DNA was collected as part of the Genographic Project, a program being conducted with the oversight of the National Geographic Society in concert with IBM and the Waitt Family Foundation. Started in 2005, the Genographic Project seeks to "collect 100,000 DNA samples from the world's remaining indigenous and traditional peoples," placing them all into a computerized database with identities expunged but sample-retrieval locations available. While the project seeks to integrate samples from more isolated, and thereby less genetically diluted, groups of people worldwide, it also welcomes interested participants to join in through a $99 contribution, providing them with their own swab kits and introductory materials.

By quickly and painlessly scraping the inside of your cheek with swabs that resemble small toothbrushes, then mailing your samples to a central laboratory for identification and addition to the database, you can add your DNA into the mix. Comparing it to the existing understanding of time-specific genetic markers and where they first bubbled into being (more about this later), scientists can get a better idea of how modern humans populated the planet.

We all know who our grandparents were; some of us even know the particulars of our great-great-great-grandparents; maybe even several generations back beyond that. But few of us know our deep ancestry. Where were the places our ancestors lived before there were things like written histories, civil birth records, and family photo albums?

No matter who you are or where you live, the DNA you carry inside every cell has already been on an astonishing trip. It's not only a trek around the world but also a trip across an enormous span of time. Quite a journey.

Science and archaeology tell us that, between three and five million years ago, the first hominids probably began roaming the African plains, standing erect on two feet for the first time. It's also estimated by some researchers that about two million years ago, an early group of humans known as *Homo erectus,* in a population that numbered a few thousand males and females, may have had a genetic makeup that paralleled that of modern humans and may have been the basis for their evolution. (The human narrative sometimes gets a little muddled, and geneticists and anthropologists are still warring over modern humankind's precise lineage.)

Still, geneticists and anthropologists do agree that some erect-walking hominids eventually became capable of making and understanding rudimentary noises or words, thanks in part to a gene called *FOXP2:* uttering sounds to denote things they saw, needed, . . . or thought. They were progressing on other levels, as well. Some time more than two and a half million years ago, they also got better at finding, making, and using simple tools that can still be found across the African and Eurasian landscapes: roughly chipped and sharpened stone knives or hand axes called Acheulean tools, plus other items, to help with what must have been a demanding—and sometimes short—personal existence.

Slowly, generation after generation, our ancestors made increasing sense of the world. Sometimes, it's estimated, no big changes happened for tens of thousands of years. Other times, as we'll see, humans and what they could accomplish through technology leapt ahead in ways impossible to have anticipated just a generation earlier.

Take fire, for example.

Fire in its wild state was probably feared at first, at least if our ancestors were intelligent enough to understand its capabilities. It burned grasslands. Beyond control, it consumed animals and people and the fruit that grew on trees—not to mention the trees themselves. It hurt you if you touched it. And if you were lucky enough to evade it, what was left behind were wastelands.

Then, through cleverness or luck, someone lost to history stumbled onto a way to, at least partially, domesticate fire. From that point, people eventually figured out how to "keep" it, providing them with useful light and warmth. Along the way, though nobody can agree on the exact location or timing, somebody probably dropped a vegetable or a piece of meat into a fire and retrieved it with a rudimentary tool—and the accidentally cooked food inexplicably smelled and tasted better. Because this "cooking" broke certain foods down a bit and probably had some effect on killing whatever bacteria might be living on it, some cooked foods also required less energy to digest. The advancement of cooked food, in a world where every calorie was needed and used for survival, proved an instant and obvious advantage.

Over thousands, or perhaps millions, of years, fire traveled from a thing to be feared to a tool to be used. Doubtless, more dozens of generations passed before someone realized that fire might potentially also be a weapon. With well-placed flames, you could burn away your rival's grasslands and game, displacing people or encircling them; threatening their lives. Conversely, by snuffing out an enemy's fires during winter, you could create cataclysm.

Fire had proven truly useful.

In the meantime, other tools were being found and developed—and used and kept. Beyond the stone knife and Acheulean hand ax, there came the thrown rock . . . the spear and the arrowhead, and the hollowed bone used as a straw or tool. Plant materials like woven palm fronds or animal skins were found to have insulating warmth. Progress on these fronts probably wasn't steady, but it continued coming. Dumb luck no doubt played a role, too, and even when one individual in a small human band happened onto an innovation, not all of the others might have been ready to adopt it. Still, day by day, our ancestors persevered. For them, the world of life-assisting and effort-saving discoveries moved forward in halting steps, each one making the world a more survivable place.

Somewhere along the way, the valleys of Africa became crowded: Roots and plants for gathering—and animals for hunting and eating—became scarce as the growing population needed more food. People began to migrate farther afield to feed themselves. More discoveries were stumbled upon, recognized, kept around. Although nobody can cite exact dates, several subspecies of bipedal hominids began to differentiate.

As I say, this took tens of thousands—if not millions—of years. But we began to get good at a few things, including gathering and hunting food. These new stores of food allowed us to get good at growing the population. We got better at spreading out. Eventually, we identified the simplest of machines. With the introduction of new technologies and understandings, a few fortunate communities leapt ahead, making remarkable advancements only to perish, leaving behind no black-box flight recorder of their crash in the form of written history.

But always, people kept going. Symbols and written language would eventually develop. So would permanent shelters, the invention of paper, and early mathematics. We continued at it, resourceful and persistent as springtime weeds. There were famines and droughts. There were ice ages roughly every 40,000 to 100,000 years. Periods of global warming came in between. But we kept surviving: spreading, adapting, and learning.

By about 11,000 years ago, humans had begun to experiment with farming. And by roughly 6,000 years ago, in a huge leap forward, people in what we now think of as the Middle East came up with a way to melt copper and tin to create bronze and, a few thousand years later, iron from rocks; in both cases employing their molten magic to improve life by making swords and digging tools and containers.

Then things got really interesting.

This is also the part of human history that probably left Mr. Anderson perched between laughter and tears. In his lifetime, he'd seen discoveries and events speed up as never before: from oxcarts to men on the moon. And while Mr. Anderson claimed not to have much in the way of wisdom, he understood that the leap in human progress across his lifetime actually carried far bigger implications. After all, he'd been on the planet for the Wright brothers' success at Kitty Hawk, the bloodbath of the Somme in World War I, the replacement of expensive, full-frontal armed assault on enemies with nuclear warfare to bring World War II to an end, and the introduction of the Salk vaccine to protect against poliovirus.

In relative terms, Mr. Anderson had seen far more technological advances than had any of his ancestors. He also understood that the velocity of the events he was seeing—even as he neared the end of his life in his hospital bed where the only technology was a small color TV bolted to the wall—was only going to increase.

Thirty-four years later, how fast is humankind moving? Consider this: Moving from the first chipped stone tools to the first man-made bronze and iron tools required about two and a half million years. From those first metal tools to this moment in which I am hurtling eastward, comfortably seated in a pressurized jet at 36,000 feet, watching in-flight movies and pondering the unilateral politics of nuclear weapons and the rise of human vaccines, took only 3,500 years.

Across these last millennia, as the accrual of human knowledge has sped up and as new ideas have been used and connected more and more readily, cultures have risen and fallen. Sometimes these cultures have prospered only to disappear (as at Machu Picchu), while others have been completely lost (such as the burial city of Genghis Khan). Other times, cultures have left behind documents and libraries whose languages we have teased back into modern understanding through discovery of records like the Dead Sea Scrolls or the Rosetta Stone.

Still, at humanity's very core, there is one library that, thanks to technological breakthroughs and recent scientific unlocking, we're just starting to understand. It lives inside each of our strands of DNA. It's the basis for how we're all the same, and for what makes each of us different, as well as a time machine telling us where our ancestors evolved and came from. All of which puts us back to where we started: pondering the Three Big Questions while confronting the Miracle at 36,000 Feet.

So let's begin again: right here, inside the cabin of a Boeing 767 in flight halfway across the Atlantic—moving at 560 miles an hour above the Earth's surface. As I sip a glass of Bordeaux, watch movie credits, and await red grapes from Peru and a tasty, supremely juicy apple from New Zealand, it's finally time to ask:

Where did we come from?

Where are we going?

And maybe most important: How did we arrive where we are right now?

*Julius Indaaya Hun/!un/!ume of the Hadzabe tribe, chief of one of Earth's last hunter-gatherer clans, near Tanzania's Lake Eyasi—a distant family member of mine*

# AFRICA

**JULIUS INDAAYA HUN/!UN/!UME,** a long-lost cousin of mine, and I are stalking the savanna of Tanzania's Great Rift Valley. It's a June morning, cool and sun-blanketed, and the sky is blue and tufted with puffy, pale-white clouds. A few miles or so to our north, the sheer, gray, 1,200-foot cliffs of the Rift Mountains and their hanging Ngorongoro Crater—which leaves the walls of the cliff looking like a broken molar—hover dramatically above the landscape like something from an Indiana Jones movie.

Closer to our feet, the earth is a powdery gray, with tufts of dry, knee-high tawny-colored grasses stretching away in every direction. Around us, scrubby acacia trees are top-heavy with green leaves silted over with gray dust. Scattered between the acacias are the hulking shapes of huge grayish brown baobab trees, their trunks 40 feet around and their limbs seemingly naked, leaving them looking like a tree's roots reaching into the sky. Taken as a whole, this place—the Hadzabe preserve inside the Ngorongoro Conservation Area—has a certain, uh, Dr. Seuss quality.

But my cousin Julius and I aren't here to sightsee. We're hunting. And with each step, we scan the underbrush and the acacia trees overhead, looking for prey we might convert into food. By reading game tracks in the soft sand and dirt, we know that several herds of delicate gazelle and bigger-bodied eland are moving stealthily and nonlinearly ahead of us through the thickets. So we're advancing as silently as possible, stepping slowly. Julius, small-boned and dark, wearing animal skins and carrying a bow and some arrows, is also regularly checking a tail of white fur attached to the base of his bow.

In the morning's puffing breeze, he monitors which direction the wind pushes the white fur's strands, which allows him to see the direction of the swirling wind, keeping us downwind of our prey and hiding our scent. Each time the breeze shifts, we turn upwind to face it.

"Do you see?" he asks, pointing at the white tuft. "You understand?"

I nod. The breeze shifts again; it now puffs from our forward left—or northwest—quarter. We turn. As we walk, Julius is also explaining to me, a big white guy in blue jeans, a button-down shirt, and moccasins, just exactly why my branch of the family left home.

"As I understand it, this place became crowded a long time ago, and not everyone could stay," he says, his language a series of clicks and pops that is incomprehensible to me (which is why we're also walking with a third guy, a Masai tribal native named Robert, who functions as my translator). "Many of our original people, many in our original family, moved away. The scientists tell me your ancestors were among those who moved. They left here and eventually populated solid ground all across the world. But now you have come back. We are pleased that you have come. You are very welcome here."

Julius looks at me. He smiles and seems genuinely glad to have me here. I arrived yesterday, having flown to Kilimanjaro International Airport from Frankfurt, then ridden a half day to this place in a car. When I arrived, I was told I'd be allowed to make camp about 300 yards from the loose group of huts occupied by Julius's familial band: a hunting-gathering clan of maybe 30, who are part of the Hadzabe, an almost extinct tribal group considered by some anthropologists to be among the oldest living African tribes, one of the two most direct links to the collective ancient ancestors of the human race.

Today, in this time of great population explosion and technological change, the land the Hadzabe live on is protected by the Tanzanian government, as the Hadzabe are also counted among the last hunter-gatherers in Africa. So permission for me to camp here was required by both the tribe and the federal government.

These days, the Hadzabe population is estimated to number fewer than 1,500. Because of their ancient heritage and their isolation, the Hadzabe's click-talking language is related to no other group, though other click-talker tribes with their own languages, such as the San Bushmen of the

Kalahari Desert in southern Africa, are considered equally old or older.

We walk silently for another minute or two. Birds chitter in the trees all around. The breeze shifts again. We turn to face it, with me not waiting to take a cue from Julius this time. Instead, feeling the wind moving slightly differently against my face, I turn into it.

Julius smiles as he sees my slight shift in orientation. "Good, good . . . yes," he says, the words coming out in his clicks and pops. He shoots me a slight, pleased smile.

Then, from somewhere to our left, Julius's pack of four small and almost hairless brown-and-white dogs begin barking excitedly, and from the densely vegetated banks of a dry riverbed off to our left, an animal squeals in fury.

Julius pauses for a moment to listen, ascertaining the direction and speed at which the prey seems to be moving, tracking the barking of the dogs and a few more of the animal's alarmed grunting squeals. In another second, Julius is selecting a few metal-tipped arrows from the passel of about a dozen he's been carrying in his right hand. He drops the rest to the ground and takes off in a loose, and amazingly fast, sprint through the surrounding brush and brambles, his strides liquid in their speed. As he streaks forward, he's dodging the larger tufts of grass, twisting his body as he keeps charging toward the noise, his right hand clutching the bow and arrows, which he holds horizontal to the ground as his arms saw the air.

As he rushes toward the gully of the dry riverbed, never missing a step, Julius turns his head slightly to the left, listening hard for the sound of his dogs chasing their prey. Then, still at a flat-out run, he shifts the bow to his left hand, and—with remarkable dexterity at a full run—begins to nock one of his arrows onto the bowstring. He's defined the angle of the dogs' pursuit, and found the dry riverbed. He's in place for an ambush. This is where he's going to meet today's lunch.

In another few seconds, the pursued, which turns out to be a shaggy, gray-black warthog, low to the ground and weighing perhaps 65 pounds, comes into sight, its legs churning as it comes around a slight bend in the dry riverbed; the coarse fur on its back puffed high in defensive fear, its eyes small and black but bright with terror.

In a single fluid motion that takes about one-and-a-half seconds, Julius halts his sprint, lifts his now armed bow, takes aim, and fires.

With a grunt and an almost resigned, low squeal, the warthog keeps running as it's smacked in the left shoulder by Julius's arrow—the arrow's contact making a hollow *whump.* The hog goes down, snout and right shoulder first, onto the powdery gray silt of the riverbed. It rolls onto its right side and jerks a couple of times, its body flexing straight and contracting, its delicately small, black-hooved feet still moving in something resembling a run. The quick paroxysms send puffs of dust into the air. The warthog goes completely still. From full run to death takes less than a minute.

As the warthog dies, the dogs streak around the bend of the dry riverbed and come upon the hog, aggressive and almost angry in their pursuit. They bark and make low growls. From behind the dying warthog, one of Julius's dogs, skinny and light brown with a blaze of white on its chest, darts in with a lightning-quick lunge, its mouth open to bite the warthog's haunches.

Julius shouts at the dog a word I don't understand but which makes the dog recoil in fear. The rest of the dogs continue to circle, but they get no closer.

That quickly, our hunt for today is over.

Two minutes later, the boxy, gray-black warthog is lying on its side, having bled its last into the Rift Valley's ash-colored soil. Julius is now smiling triumphantly. The cane shaft of his biggest arrow, its head made from nails Julius heated to bright red in a campfire then pounded into an aerodynamic dagger, stands tall in the air, sticking from the hog where it entered, just behind the animal's left shoulder. "I got him in the heart," Julius says, clicking and smiling as he explains. "That is why he fell immediately to die."

As we stand and watch, other male members of my extended family are now arriving on the bank of this dry river, having followed the dogs' barking and the squealing hubbub. They're running through the savanna underbrush, coming closer. As they arrive, seeing the warthog in the dirt, they hop up and down and talk in a forceful mix of singing and clicks. They're smiling and dancing and bouncing their fists upward into the still-cool blue of the morning sky.

Without anyone saying a word, one of the tribesmen pulls out a Bowie knife and clears a circle of dried grass from the savanna, creating a six-foot circle in the dust maybe 50 yards from the dry riverbed's deep U-shape. Then, he lays the knife on the ground in the center of the circle as another tribesman extracts a narrow, four-inch length of pale wood from a rough leather bag slung over his shoulder. The wood has sockets bored into it. The two men set the wood over the knife blade to stabilize it on the soil, and, finding a three-foot cane of dry brown reed and some bits of dry grass and tindery tree bark, they crush the grass and bark between their hands then drizzle it into one of the sockets. Now it's time to start. One of the men sets the base of the cane into the same socket as the crushed grasses and begins to spin the cane back and forth quickly, rubbing it between his palms.

They are making fire.

In less than a minute, the first whiffs of pale-gray smoke lift from the socket. As more smoke rises, the men add slightly larger bits of tree bark, puffing on the new flame with light breaths, their lips pursed: it looks almost as if they're kissing it. Atop the socket, they add hand-size stalks of dry grass, which have been crushed inside their hands to create an airy, fire-friendly nest. With the addition of the dry grasses, the first orange fire flares into being. Accompanying the arrival of actual flame, the man with the knife lifts it and the wooden block away, dumping the fire and new embers onto the earth. They keep blowing on the burning pile of tinder, ever so gently. They add another nest of grass and then some small twigs to the flame, then larger twigs . . . then branches. From start to full and blazing success takes about three minutes.

We now have a cooking fire going, so the knife can be put to other uses.

Back down in the dry riverbed, Julius and a few of the other tribal men have been keeping watch over the kill. When it's finally determined the hog is dead, following a few gentle prods with their sandaled feet, the tribesmen again use their feet to roll the animal carefully onto its back as Julius extracts his arrow from its left side with a single firm tug, like a tailor popping a stray thread from a hem. As this happens, another man keeps a knife near the animal's throat, in case there's a last, slashing lunge.

In two quick, deft motions, one of the tribesmen deeply cuts the arteries of the warthog's throat. Very little new blood flows. We wait another minute. The warthog is now ready to be lifted and carried nearer the fire for preparation and cooking. Two of the men lift it and walk it up the bank of the dry riverbed, where a bed of freshly stripped green leaves has been laid out on the ground a few feet from the flames. The pig is set onto its back.

"I'll return in a moment," Julius says. "I'm going to retrieve my other arrows." Leaving his bow and a few arrows on the ground next to the hog, he trots off, back into the savanna as I watch the other men clean and dress the warthog.

Spreading its front legs, one of the tribesmen cuts the hog's front legs and shoulders free, dropping the legs onto the leaves. They open its belly, popping the diaphragm and peeling it away before removing the internal organs on each side of it, the heart and lungs and viscera. They clean some of these organs, the kidneys and heart, on the leaves while hanging others (the liver, the stomach and intestines) on tree limbs. After that, they scrape out the body cavity, working the knife gently against it, until it is smooth and very clean. The whole process takes about ten minutes.

When the thorax is immaculate, the man with the knife slices along either side of the spine, removing the ribs and using the knife to separate them from the skin once the bones have been popped free. The remaining carcass—head, snout, haunches, offal, and fatty back—is then lifted away and draped over a low tree limb. This is reserved for the clan's women, children, and elderly. But the rest of it? The kidneys, heart, front legs, and ribs? We're going to eat that meat here, right now.

Julius returns, carrying his jumble of arrows; he puts these in a pile with his bow and other arrows, counting them all once to assure himself that everything is together. The fire takes about 15 minutes to cook down, during which time we stand around, talking a bit, but more often silent in anticipation.

Julius stays busy, pointing out a vial of poison that he carries in a leather container and uses to coat some of the arrowheads he uses on bigger game. "I shot a wild water buffalo once with this bow and arrows," he says. "It didn't take ten minutes for that buffalo to die. I followed it, and then it died. This poison is very powerful. It comes from a plant, and I will show that plant to you another day. But, as we teach our children, you need to be very careful around this poison once it is ready to be used. You cannot get it in a cut on your hand. This is very serious."

Julius finishes his coating of the arrowhead. The fire burns lower. All the men—there are maybe 12 of us—remain expectant and happy; there's little need for conversation. A few of them regularly slide their hands together in anticipation of the meal. Like many hunting-gathering groups (the San Bushmen included), found or foraged food is treated communally. "This is how it has always been for us," Julius says. "We share. You will come to see it. There is always enough."

With the hog now cleaned and finished, the man who wielded the knife cleans its blade, first by wiping it with his fingers, then scrubbing it lightly with sand, then wrapping it with large leaves stripped from a nearby tree, and sliding the leaves down the length of the steel, leaving the blade spotless.

When the flames have died down and the bright-red coals are whiskered with white ash, the hog's ribs and still hairy front legs are dropped directly into the fire. Predictably, the smell of singed hair fills the air.

Soon, the meat in the fire is crackling and hissing. It smells good. Julius is smiling. "So now you know . . . you have seen it. This is how a day goes here," he says. "Because you are my guest, I'd like you to go over and point to the meat you would like to eat. Are you hungry? The food will be ready soon."

The moment brings me to a strange emotional crossroads. There's certainly a giddy, abstract sense of a hunt's success: of completing a job and having it come out better than expected. But, for me, that's mixed with a very real and disorienting horror as we cook the thing that was just killed and gutted, the pool of its thickened, darkening blood still sinking into the dusty earth. And yet, I also know this is one of the reasons I climbed aboard that United Airlines flight back at Washington Dulles.

Fact is, I've been dreaming about this exact scenario for months. And now here it is . . . and I'm not sure how to deal with it.

Finally, I step over to the fire and point into the thick gray smoke. "I guess I'd like to eat a rib and a bit of shoulder?"

Though my cousins have been doing this every day for about 60,000 years, I've just put in an order for my first home-cooked meal, Hadzabe style. So while it's strange to me, the Hadzabe think nothing of it. Nobody even gives a sideways glance.

Fifteen minutes pass. Eventually the men employ sticks to lift and pull the warthog's legs and ribs from the fire. They lay all the cooked meat on an enormous green leaf, two feet across, which was pulled off a nearby tree. The meat is juicy and bubbling, though caked in spots with white ash. The men use the knife to further carve up our feast. Everyone gets a rib. Some wipe the ashes off with their fingers; others simply eat the meat, ashes and all. The ribs are crunchy outside; inside, they are juicy, though strong tasting. Inside the cooked meat flavor, there are hints of ammonia and nuts.

The shoulder is carved: Layers are pared away lengthwise from the warthog's robust upper shoulder joint down toward the hoof. Each slice comes off in a rough-grained strip, like chunky pieces of steak. There's less ammonia in its flavor. The edges of each slice are browned to a warm, charred crunch that's surprisingly satisfying.

As we finish the meal, each of the men signals he's finished by crouching down, placing his hands in the sandy silt, and "washing" them in the dust. Then he stands again, rubbing off the last bits of earth, hands clean and dry.

As I crouch to wash my hands in the dirt, I realize that, somehow, this has been a remarkably satisfying meal. I'm satiated, if not really full. Still, I don't want to eat any more, even though another portion is offered. Truth be told, I've only had bites, as I don't want to deprive anyone of needed food. But there's something else going on, too. In a human tradition that stretches back far beyond when records of such things were kept, having now taken a meal with these people—having "broken warthog" with the Hadzabe—I feel somehow closer to them. I feel more like one of them.

Over the last hour, the day has grown hot, with the June sun staring down from near its apex in the sky. In his click-talking tribal tongue, Julius tells the other men to gather a few long sticks. With these, we throw together a makeshift rack, and begin to carry the remains of the warthog back to camp, where the women will cook up the rest of it for the group.

"When we get back," Julius says to me, "the women and children and old people will eat. They boil their meat, which I believe isn't as nutritious . . . and it doesn't taste as good. But it is easier for some of them to eat that way. For us, it will be time for a midday rest. Then we will make a nice afternoon."

Never underestimate the joys of reuniting with family.

Here's a confession we all must make occasionally: I woke one morning to discover that many of the things I had assumed about my life were wrong.

Like everyone in my family on both my parents' sides, I knew we'd been American Midwesterners for generations: Ohio land-grant farmers and Illinois rural doctors since before either one of those places had officially received statehood. Before that, we were old-line, largely Anglo-Saxon immigrants to America, who'd been washed upon the New World's shore during the first wave of European settlement. Fact is, I can trace many of my direct ancestors' arrival in the United States to before 1635.

My favorite ancestor is a shipwrecked Scottish sailor named Glann who (though my mother grimaces every time I suggest this) might have even been a slave-ship owner and captain. He arrived in America in 1633, after the salt-laden ship he is said to have co-owned was sailing back to Liverpool from the Turks Islands in the Caribbean and foundered in a horrendous storm. After drifting at sea for weeks in the hull of his vessel, he was picked up and put ashore at his rescue ship's next port of call: New York.

He never made it home. Instead, he started anew in the American Colonies. "He soon came to the conclusion that he had seen enough of the vicissitudes of a seafaring life," or so the family history goes. "He pushed toward the country in quest of work. At Kingsbridge he fell in with a Dutch farmer who set him to threshing, and he wielded the flail with such energy and success that he got a permanent job. After three years, he married a daughter of his employer, and continued to work for his father-in-law until he was able to run a farm on his own account."

Eventually his descendants would own the farm that became Yonkers, New York.

Another of my ancestors, Matthias Farnsworth, was a weaver of fine cloth and a Puritan pillar of society in what would become Lynn, Massachusetts, part of John Winthrop's Massachusetts Bay Colony. Records indicate that he sailed from Lancashire in England in the early 1630s, though there is no public accounting of him in Massachusetts until the 1650s. Still, by then he had married and had several children and had been in the neighborhood for some time. And remember: all of this happened long before George Washington was a half-pint in short pants, said to be toting a hatchet somewhere in Virginia.

Then there's Pietro Cesare Alberto. A son of Italy and a sailor out of Venice, Pietro fetched up as an immigrant in New Amsterdam in 1636—my family's slow-boater—following a long passage to the New World. Still, with his name newly Anglicized to Alburtis, he worked his way into the fabric of the New World, too, with his son eventually marrying into the Bogardus family, of which the most famous early American member was New York's Everardus Bogardus, referred to as "a stout, hard-drinking Calvinist minister" in Russell Shoro's masterful book of proto-Manhattan history, *The Island at the Center of the World.*

These were the stories I carried around as my family's history. And like everyone in my Illinois-, Ohio-, and Wisconsin-based clan, I was comfortable with these intertwining narratives that, inexplicably and yet inevitably, led to the lot of us. We farmed and remained in the clergy and became frontier doctors. Locking in these certainties were my mother and my Uncle John Burchfield (my mother's younger brother) and my Uncle Steve

Webster (my father's brother): who for the last 25 years have devoted months and months of research and travel to documenting all this. From the meetinghouse records of William Penn's Quakers to the death records of Skowhegan, Maine, to the deeds of land at Lucas County, Ohio, they have been so tirelessly diligent that today, after years of poring over transcripts of oral histories and partial documents, my parents now possess an office full of old tintypes and photographs, family-history books, and hanging folders fattened with layer upon layer of age-browned and crumbling documentary paperwork.

As a family, my parents, sister and brother, uncles, aunts, and cousins have our heritage pinned fairly neatly down, at least as much as many Americans do: former immigrants inside a nation of immigrants. We're covered. And although I wouldn't have admitted it as a kid, when my mother first began looking into all this, over time I've discovered a certain warm comfort in knowing the deeper facts of my family's past. This only became truer when I met the woman I married, her family largely Scots, many of whom arrived in the United States about the same time my ancestors did, and we had our own children to carry the family history forward.

My past was orderly. Comprehensible. Where my life sat made sense.

Then Dr. Spencer Wells dropped into my life. A cool guy with a Harvard Ph.D. in population genetics, Dr. Wells, as part of a worldwide team of mentors, assistants, and colleagues, has spent the past two decades traveling the world, taking blood or cheek-scrape DNA samples from hundreds of people from somewhat isolated, and thereby more genetically homogeneous, cultures.

Building on the work of others across the century before him, his work has been assisted by computer-generated science, which can examine and prove new genetic findings with increasing accuracy and speed. Wells gets this. "As often happens in science," he is famous for saying, "technology has opened up a field to new ways of answering old questions, often providing startling answers."

He's right about that, too. Over the last decade, using digitally based science and computational speed, Wells and his team and colleagues have bounded ahead, gaining new understanding of the genetic origins of humankind, as well as the age of modern humans as a species. It's helped to show Wells and his colleagues how human populations have migrated around the globe, inhabiting the entire planet from an original tribe that, as Wells says, "might have originally numbered as few as 2,000."

When I visit him in his office at National Geographic Society headquarters in Washington, D.C., a windowed room strewn with a world of cultural artifacts (from a Hadzabe hunter's bow to a leather Mongolian jacket tossed over the back of a chair to the small, black-rubber orb of a double-yellow-dot squash ball on his desk), Wells offers to walk me through the progression of population genetics findings slowly.

Friendly and lash-smart, Wells is a ruddy-complexioned speed-talker with an obvious mastery of his subject. He grew up in Texas, studying at the University of Texas by the age of 16 and receiving his Ph.D. from Harvard by 25. Still, his demeanor and mannerisms aren't really tethered to either location. But, then, he did postdoctoral work at Stanford, followed by an extended stint as a researcher at Oxford University. Field seasons have taken him to Africa and the plains of Central Asia. He is literally a man of the world.

He is also, it goes almost without noting, a respected genetic scientist with a global platform and a reputation for fast-track accomplishment. Just last year, after a handful of TV documentaries, more than 40 published journal articles, and two books, Wells finally turned 40: a fact that seems to have surprised even him.

Still, he doesn't dwell on it. But then, he doesn't dwell on much. In fact, once he's on his favorite of subjects, the creation of computational models to study population genetics, you quickly notice that Wells's mind has often moved to its next point as he's still finishing the sentences underlying his last one, his thoughts and ideas chugging relentlessly forward.

In the interest of my better understanding his project, however, Wells offered to slow down, schooling me at a pace that I could accommodate. And wanting me to have a better understanding of how his methodology has come to exist, he fired up a PowerPoint presentation on his laptop and began his explanation by taking me back almost a century.

"Let's start with the very first study of genetic anthropology, by Hirszfeld and Hirszfeld in 1919," he says. Published in the British medical journal *The Lancet* under the title "Serological

Differences Between the Blood of Different Races," the results of research on the Macedonian front, it was the first scientific study of genetic anthropology. Discovered almost accidentally, the study's roots were in the research of two married scientists, Ludwik and Hannah Hirszfeld, who began questioning whether different blood groups were or were not related to geography. Blood typing itself had been discovered about 15 years earlier, in 1901, by Karl Landsteiner, who noticed that mixing the different blood of unrelated individuals sometimes left the blood coagulating and clumping together unnaturally. Before long, Landsteiner had identified what would become the first two blood types: A and B (types AB and O would be discovered later).

"So the Hirszfelds were working during World War I," Wells says, "and obviously, as it was World War I, they were giving a lot of transfusions. As they were doing this, they began to notice all these interesting patterns. They looked at where the different troops they were transfusing came from, and they noticed that the incidence of type A blood increased in troops who came from farther in northwestern Europe, while people possessing blood type B decreased. They began asking: Why is that? So they decided maybe there could be two different origins for humanity."

The Hirszfelds hypothesized that all humans were divided into two separate biochemical races, A and B, and because of the polarized saturation of these blood types in different areas of the globe, the two groups must have had different geographic origins. Because blood type B increased in frequency through populations trending southeast across Europe until it was overwhelmingly found in Allied troops from the Indian subcontinent, the Hirszfelds surmised: "we should look to India for the cradle of one part of humanity." By contrast, northern Europeans appeared to be a source of type A humanity. The Hirszfelds then hypothesized that, over time, "a broad stream of Indians passed out [of their traditional homeland area] ever lessening in its flow, which finally penetrated Western Europe." This accounted for the existence of some type B blood in lessening but existent amounts in northwestern Europe, although the farther northwest you went, the lower the frequency became.

Groundbreaking as the idea was, it was also proof that making global population-model theories from limited sets of data can be a dangerous thing. Still, as a theory, the Hirszfelds' article was an early building block in the concept that human biological differentiation existed on a cellular level, far deeper than was externally visible. It was the first proof that, thanks to deeper genetic differences, people in different areas around the world had probably migrated over time from other locations across the planet's surface.

But after several decades of general acceptance of the Hirszfelds' theory, that different blood types had their origins in different areas of the world, and that blood type was an indicator of origin, their theory was turned on its head. This was accomplished by the then Chicago-based statistician and college professor Richard Lewontin. In the 1960s, Lewontin had been working in fruit-fly genetics, studying among other things the variety of cellular proteins made by these insects, when he began to wonder if humans of every race made identical kinds of proteins inside their cells; materials themselves created by DNA.

Back then, science wasn't yet able to break down variations inside DNA in any understandable way. But Lewontin knew, as did most scientists, that DNA was likely responsible for replicating itself and creating cellular proteins. So, he theorized, if he were able to assess the differences in the variety of proteins created by the DNA-based instructions in our genetic code, might that provide a glimpse into whether we are all one species with identical DNA bases, with only small external differences in things like hair color and skin pigmentation? What would the ratio of similar to different proteins be in similar and different human populations around the world? If that ratio was broad enough, did that mean we were multiple subspecies inside the single species of modern humans known as *Homo sapiens*? Or, if the percentage of similarity was high, were we all simply a single species with external physical variations? How much variation could there be in the cellular proteins of different races and peoples?

So Lewontin set about testing human proteins, and what he discovered was astounding: 85 percent of the roughly 200,000 different proteins created inside the cells of all the world's people are exactly the same. He also learned that variations of the other proteins do exist between different tribes and peoples, but that inside these smaller groups, another 7 percent of the proteins were also the same. Still, the take-home message was unavoidable: no matter what population you sampled, Koreans or Hadzabes; Parisian French or the Yanomamo natives in southern Venezuela; blood type A or B or AB or O; the proteins made in the cells of all the world's people were overwhelmingly the same.

This finding also meant that we all—every one of us—carry 85 percent of the world's genetic diversity in each of our cells; in each of our collections of DNA. This was the first hint that, rather than evolving separately and independently as different races in scattered places around the globe, perhaps we all came from the same source.

"And that's where Luca comes in," says Wells.

In the 1950s, just a few years before Lewontin began to ponder the similarity between human proteins in different cultural and tribal groups, a population geneticist and physician named Luca Cavalli-Sforza (an Italian by birth and education who later worked at the University of Cambridge) began wondering if he sampled different populations and statistically broke them down by blood groups, A, B, AB, and O, plus their various subtypes, would the blood groups exist in roughly the same frequency everywhere, or would there be different ratios of blood typing among different races and cultures?

As he began summing up the frequencies of different blood groups inside their own units and against different populations, Cavalli-Sforza discovered that some cultures, often those living in close proximity to one another, often possessed statistical similarities. As a visual tool, he started grouping these different peoples and tribes on paper, by culture and geographic region.

If clusters of statistical frequencies of blood type in different island populations in South Pacific were similar, for example, these were connected on his diagram by short lines, whose length was determined by the percentage of similarity. Those populations with less statistical agreement were separated from one another by longer lines on Cavalli-Sforza's chart. As he kept working back toward lower and lower percentages of blood type frequencies, the lines grew longer and longer, until Cavalli-Sforza's diagram looked unmistakably like a branching phylogenetic tree that had been laid on its side, with a single ancestry eventually being found on the tree's "trunk," and with the various cultures spreading upward and outward from there.

"So that's where we were," Wells says, "we had a tree showing how similar some populations are, based on the relative frequencies of their blood types, and eventually that tree implied an eventual lineage traced back to a small group of ancestors . . . and that made a certain amount of sense. The problem was, there was no estimate of timing on this. We didn't know when across time these populations shared a common ancestor, or when they diverged from each other. But that was also soon to change, since the technology we were using to assess this was beginning to change, too."

By the early 1970s, Cavalli-Sforza had moved to Stanford University. There, armed with early computers, he and another scientist, Dr. Allen Wilson, who worked across San Francisco Bay at the University of California in Berkeley, began to drill a little deeper inside the constituents of our cells, slowly coming to see what's contained inside any individual's DNA sequence.

By this time, early computer analysis of actual strands of DNA was becoming possible (it would take until the early 1980s for reliable and fast sequencing to become a reality), with Cavalli-Sforza and Wilson leading the charge.

Of course, even arriving in the age of early genetic sequencing had been a century-long saga. DNA had first been isolated in 1869 by the Swiss physician Friedrich Miescher, who'd discovered unusual proteins inside cells while studying discarded surgical bandages under a microscope. Miescher's findings—much like the rise of life itself—took some time to gain traction. In fact, it took him 23 years to arrive at the conclusion that, just maybe, the compound he called nuclein (because it lived in the nucleus of a cell), might be how hereditary characteristics were passed from parent to offspring.

Then, in 1919, a Russian-born biochemist named Phoebus Levene, working at the Rockefeller Institute in New York City, found through enormous trial and error that all the nucleotide bundles inside the cell's nucleus consisted of only four kinds of sugar: adenine (A), guanine (G), cytosine (C), and thymine (T). These, he believed, repeated themselves over and over in sequence. Might this repeating sequence be how DNA encoded the blueprint for all life on Earth? Might that code prompt cells to make more and more identical cells, each differentiated by instructions given by this long, simple chain? Still, it all seemed too simple. Could the combination of four different sugars really dictate the characteristics of all of the varied life on Earth?

It took most of the next 50 years of science and discovery to prove, but the answer was yes. Science found that DNA existed in long, unique, double helix chains of paired molecules in combination, each of which, in humans, is broken into 46 individual chromosomes that occur inside each of our cells: 23 of them from our mother, and 23 from our father.

And they are long chains, even the shorter versions of them. Each strand of DNA contains some 3.2 billion individual sites for A, G, C, and T nucleotides, which is enough to combine for a possible 10 to the three-*billionth* power of available combinations. Given the diversity of these combinations, it's a virtual guarantee that every person who has ever walked the Earth has also been a unique DNA sample in the world.

Small as it is, however, the enormity of DNA inside each cell, and its characteristics, can be spun in other mind-blowing directions, too. Taken as a whole, each of our cells somehow manages to carry a little more than six feet of unique DNA inside its walls. Geneticists like Wells estimate that, were you to take all of the DNA contained inside every cell in your body and connect it into one long and microscopically helical chain, the combined length would stretch to the moon and back more than 3,000 times.

Yet despite its simplicity in combinations, and because of its length and the fact that DNA does nothing but replicate in our cells nonstop, it is also prone to small mistakes in its transcription sequencing: glitches called mutations. In most people's DNA chain, in fact, there are as many as 100,000 of these tiny transcription errors, and they can occur naturally or as the result of external influences like electromagnetic or nuclear radiation, or even extreme heat. Still, although some mutations can cause things like cancers and disease, most are harmless.

"So back in the late 1970s and 1980s," Wells says, "as science became more and more capable of reading the sequences on the human genome, thanks to advances in technology, we found that, inside each of us, we have the main sequence, which is basically the same for everybody. Along the DNA chain, there were various insertions and deletions—an A was mistakenly put in for a T in one spot or another—but now, thanks to computer mapping of each individual site on a DNA molecule, we were able to track them all. So at a particular site number, which is the same in everyone's genome, where we'd expect to find the same nucleotide, an A or a G or whatever, we saw that this gene has a mutation at that marker. By testing lots of genes over time, we'd also come to learn that DNA mutates at a fairly measurable rate. About 100 new mutations occur per generation, which meant that genetics also had a built-in clock. And from our perspective, this is where things started to get really interesting."

In each living complex organism, when its two parents' genetic characteristics come together to create a new baby, whether a human, a horse, a bald eagle, a staghorn coral, or a bluefin tuna, each parent contributes half the total number of chromosomes, the male through its sperm and the female through the egg. Once united in the fertilized egg, the DNA strands then recombine in a slightly "shuffled" form, the mechanics of which scientists have yet to completely understand. Still, this shuffling creates an entirely new genetic code, yet one with specific echoes of those genes that came before.

In the human case, with 23 new chromosomes coming from each parent, the 50:50 split of parental chromosomes in a new baby's cells creates both the mixed parental characteristics of the child—maybe it has the iron stomach of its father, the eyes of its mother—but all of it is now contained in these new, shuffled DNA sequences. This is one way that genetics both evolves coherently and yet retains its remarkable diversity over time.

In humans, this shuffling takes place across the entire genetic spectrum, with the exception of two areas. And in these two areas, genetic codes are passed down intact from generation to generation. This, when paired with the "genetic clock" contained in each strand of DNA's mutations, means that if science had a way of accurately counting and collating these mutational changes, say, inside a huge computerized database, it could work backward over time, figuring out aspects between people such as common ancestry.

In males, the stable DNA sequence is passed along by the father's sex-determining Y chromosome: the chromosome that pairs with the X chromosome donated from the mother to determine male sexual characteristics, never mixing, recombining, or shuffling with any other DNA. Every Y chromosome is passed down virtually intact from father to son to son . . . to son.

In females, the X chromosome does recombine and change over time and so is not passed on intact from generation to generation. It was discovered, however, that a small ring of

DNA exists outside the nucleus in the female cell's energy-producing mitochondria. Called the mitochondrial DNA, or mtDNA, it is passed down identically from mother to children of both sexes, tracing a maternal line of ancestry. Because only women pass mtDNA along to their children, and because mtDNA never interacts or recombines with any other DNA in a cell's nucleus, geneticists discovered a stable DNA sequence in women, as well.

It was the discovery geneticists had sought and awaited for a century.

"So we now had stable genetic sequences to follow in both men and women, plus the known markers of the little genetic clock to follow," Wells says. "We had a powerful tool to determine when different markers arose. Then, by sampling a variety of people and indigenous tribes around the world, and tracing backward to common markers, we could figure out who shared those markers. We could start to see who shared common ancestries."

As statistical samples of genetic markers grew larger and larger, Cavalli-Sforza and Wilson and their colleagues were better able to see how these markers were present in some areas where all of the inhabitants shared the marker, with statistical patterns of fewer and fewer people possessing the marker radiating away from that marker's population center in a starburst pattern. And with each new collection of DNA poured into the computer models, human migration patterns began to reveal themselves.

As fate would have it, when Wells arrived at Stanford for his postdoctoral work, he soon became bewitched by the computer-assisted assessment of genetic markers and the migration patterns that Cavalli-Sforza was revealing. Already, Cavalli-Sforza, who was concentrating on the lineage of the Y chromosome, and Wilson, who was uncovering the heredity of mtDNA, were beginning to make sense of DNA's history through new computer-driven tools and methods of deeper assessment. "That was so interesting to me," Wells says. "The work they were doing blew my mind. There was so much potential there to employ science and new technology, to really learn more about our common past, that, well, I just had to get involved. It was the opportunity of a lifetime."

Soon Wells and his colleagues were traveling the world, taking blood and cheek-swab samples and searching for marker mutations on the stable Y chromosomes and mtDNA bits in each new individual or group. And just as Cavalli-Sforza had done with his earlier blood type studies, they began to draw a sideways tree of genetic markers. Starting far out at the "leaves" with today's disparate tribes and cultures, they searched for common mutational markers that, like the branches holding the leaves, would show common ancestry.

These male and female DNA marker commonalities became known as haplogroups, from the Greek *haplo,* meaning "single" or "simple," and the shared common ancestries began to stretch across the globe. Cavalli-Sforza's diagram united very different, distinct groups of people through their unique mutational genetic history: a commonality that could exist no other way.

As haplogroups came together, they were given capital letter designations for larger groups, with lowercase letter designations based on smaller shared traits. These lowercase letters are often followed by number designations to denote when each new mutation was found in the sequence.

Tracing mutations backward to more and more commonly shared genetic markers, the Stanford and Berkeley teams also began to see something that anthropologists had long suspected but previously had no way to prove. And Wilson's team found it first. Through the study of mtDNA, it could be shown that all women shared a single, common female ancestor, predictably called "mitochondrial Eve," in Africa. And because of the genetic testing the teams had done in that region, it could also be shown that she probably had originated somewhere near the Rift Valley in eastern Africa, or in the Namibia area farther to the southwest: the first female ancestor of today's modern humans probably arose about 200,000 years ago.

Soon, Cavalli-Sforza's team had tracked their "Adam" back, as well—to sometime around 60,000 years ago. (As it turned out, Y chromosome DNA is harder to chase deep into history.)

Eventually the teams developed and published their findings, with the announcement of a common "Eve" making the front page of the *New York Times*. The teams were subsequently hailed as the scientists who'd finally cracked wide open the mystery of human origins. With their success, the teams first expanded and broadened their study, only to see some of their members then drawn away to different institutions to continue their work.

In Wells's case, he continued to do field science on the deep genetics of people in one of the world's most genetically diverse groups, Central Asia. In 1999, he took his research to England and the University of Oxford. While at Oxford, he was involved in a Channel Four documentary that led to his first book on the subject, *The Journey of Man,* which also became a PBS/National Geographic television special in 2002.

"From there," Wells says, "things sort of took off. National Geographic and I began talking about a project that might take the findings of deep ancestry and expand them. And in 2003, I became a fellow at the National Geographic Society and began to work on the materials and science that would ultimately become the Genographic Project."

As it happened, early in the life of the Genographic Project, I was among the first to deliver my DNA to be run through Spencer Wells's computers. My plan was to identify my distant ancestors and then visit them, so that we might become acquainted.

But, truth be told, my real intention was far simpler than that: I was curious. I wanted to look those ancestors in the eye.

Yet if heredity is fate, then Spencer Wells and the Genographic Project were about to deliver a fateful shock. Instead of reunions filled with light rain, warmish beer, grouse shooting, sailing, and moody bouts of salmon fishing, with a sunny Italian side dish of amarone and prosciutto tossed in, my DNA-based results told me I was not only kin to Julius's click-talking Hadzabe bushmen in Tanzania (among the closest living links to genetic "Adam"), but also to Lebanese Arabs, tribal Uzbeks in Central Asia, and Spanish Basques.

In fact, my Certificate of Y chromosome testing and the accompanying map showed my haplogroup as *R1b* (*M343*), one of the oldest out-of-Africa lineages possible; the arm of the human family that, according to Wells, first populated Europe more than 30,000 years ago. Or, as Wells put it in his book *Deep Ancestry:* "These travelers are direct descendants of the people who dominated the human expansion into Europe, the Cro-Magnon. The Cro-Magnon created the famous cave paintings found in southern France, providing archaeological evidence of a blossoming of artistic skills as people moved into Europe."

Along with my Genographic Project certificate came a map that every Genographic member gets, customized to show my individual genetic path. It was titled "Migration Routes: Donovan Webster," and it showed, through common DNA markers in populations across the world, how my particular genetic code had moved out of the Rift Valley, labeled "Eurasian Adam: 31,000 to 79,000 years ago," then across the Arabian Peninsula, into the Middle East, and, by about 45,000 years ago, into the Asian plains perhaps 1,000 miles north of what is modern-day India. From there, my genetic route made a huge, sweeping turn west: through central Europe and into Basque Spain in the north of that nation.

Knowing what I did of my family history, none of this made sense. My DNA came out of Great Britain and Venice, didn't it? When I called Wells to double-check my results, he was straightforward. "What part of Britain does your family come from?" he asked. "Your genetic route is virtually identical to a lot of Anglo people's. See, 10,000 years ago glacial ice still covered northern Europe, and people with your, and my, genetic makeup were living along the glacier's southern boundary, largely in what today is northern Spain. As the glaciers receded, our ancestors followed them north, eventually arriving in the British Isles."

Wells went on to tell me that, along with everyone else's, my genetic material started out in Africa between 100,000 and 200,000 years ago. But it's believed that as pickings around the Rift Valley grew scarce in the face of overpopulation or environmental change, members of my *R1b* genetic group migrated north into the Middle East, beginning about 65,000 years ago (though, according to Wells, my particular markers probably departed Africa closer to 50,000 years ago). From there, the human genetic tree splits: some people's DNA tracked west toward Greece or east for Iraq. Mine, however, went northeast, over the Caucasus Mountains and into Central Asia. Finally, 10,000 to 30,000 years back, my DNA turned west and followed the glacial ice sheet's southern edge to the green hills of what is today Spain.

As we spoke on the phone, Wells's words washed past like a rushing river. Africa? Uzbekistan? Spain? Come on. What he was talking about seemed impossible. But, as Wells was quick to point out, given that the computers had 100 percent identified my genetic history and its unique markers, each trapped forever in the same place inside my DNA like donuts slid along a string, there was no other way these markers could have gotten into me. I was matched to these other genetic groups. Lancashire and Venice were other bus stops on the way to today. My heredity was my deep history.

Within weeks of receiving my Genographic report, and with Wells's assistance in isolating tribes and individuals to whom I was related, I pieced together travel plans to put me in contact with my genetic family. As the route became clear, I grew excited. Despite having traveled to more than 100 countries across a 20-year career as an author and news correspondent, I'd be seeing parts of the planet that I'd never experienced before.

I mean, did I really have genetic relatives in Uzbekistan?

And so, on this June afternoon, it came time to board that United Airlines jet at Washington Dulles International Airport for my trip back though genetic time, starting with Julius and, a month or six weeks later, ending in the Basque homeland of northern Spain.

The beginning of any journey is exciting. Still, despite having to come to grips with my new deep history, there remains something impossible to envision about it all. I am going to places I've never been before, where I know no one. Yet I'm headed there to meet members of my ancient tribal family; to journey to their homes and live with them and eat with them and get to know them as they live their own lives in places where my DNA had once been, but where I'd never made a footfall. What will the experience show me about these extended members of my family? And what will my distant relatives show me about myself?

After all, I'm just related to them . . . and they are related to me. And we have a lot of catching up to do.

---

Three airplane flights and a day's drive later, I am ready for my first "reunion." An hour after I arrive and make camp beneath a spreading acacia tree on the Hadzabe reserve, Julius strolls over from his settlement to say hello for the first time. He's heard my Land Cruiser pull in, and, as Spencer Wells sent word ahead, he's ready to visit. As he walks closer, my sense that there is no way he and I could be related grows only stronger. "Basically, yes, you share markers with Julius," Wells told me before I left. In my case, it was the *M168* marker: one that's extremely common among men. "But then," Wells continued, putting a finer gloss on it, "the Hadzabe share markers with almost everyone on Earth. They have the most diverse genetics on the planet. Everyone can trace their DNA back to the Hadza and the San bushmen. They're where the trail forward in time begins."

Julius is dressed in a tunic of tanned and sewn-together animal skins. On his feet are sandals with uppers of leather and soles made from the treads of car tires. Around his head, he wears a brightly beaded headband, a single feather emerging from its center, on his forehead. As he approaches, he's tentative. He moves cautiously. The closer Julius gets to my camp, the more I can't imagine anyone with whom I have less in common. He may have the *M168* marker, but everything outward about him is different from everything outward about me.

When Julius arrives in the grove of acacia trees, Robert, my Masai interpreter, makes introductions, and before long we are sitting in wooden folding chairs in the shade, each of us holding a bottle of cold water from my ice-filled cooler, getting to know one another. At first, it's slow. But as we begin talking, it's the things we share that become more obvious. We both have several children under the age of 12. We are both mindful about our own future, and the future of our families.

In my case, I'm saving for my kids' education with, I hope, something left over for my wife and me in old age. In Julius's situation, the hunter-gatherer Hadzabe people are being encroached upon by several pastoral and seminomadic tribes, particularly the Datoga and the Masai, whose animals are now grazing illegally on traditional Hadzabe lands. The grazing has limited what there is for the hunting-and-gathering Hadzabe themselves to eat; the livestock's pressure on the land seems to be driving off some of the precious game Julius's people need to live.

"I am worried," he says in clicks and pops. "We don't keep animals. We are hunters and gatherers, and if the wild animals and food are taken from us by grazing by others, our people will not have enough to survive. This is serious. I have recently gone to the authorities, and they say that they will help, but we have not really seen that assistance yet. Still, they have always been good to us, the government has always looked out for the Hadzabe in the past, and so I hope they will do this again now."

Julius and I have other mutual imperatives, too. As hunter-gatherers, if the Hadzabe get smacked by the flu or some debilitating virus or disease, nobody eats. It's the same for me. As an author and freelance magazine writer, there's a sometimes scary cause-and-effect relationship between stories written and chow on the table or gas in the car.

When I tell this to Julius, who has been away from the Hadzabe lands to interact with government leaders, and who flew from Tanzania to Washington, D.C., to be present at the initial announcement of the Genographic Project at National Geographic headquarters in 2005, he smiles.

"Yes," he says. "In the end, all men are somewhat the same. We work to live on this Earth, so there must be some similarities. That has always been the way."

I've done some homework on the Hadzabe and have learned that they're matrilocal: meaning a man takes one wife, then moves in with the wife's family to live. When I ask what it's like to bunk full-time with the in-laws, Julius smiles again. "Men are still important among the Hadzabe," he says. "And men are the tribal leaders. Since my teenage years, I have also been the Hadzabe chief. It is a job that was given to me by the elders, because of my abilities. It is my job to lead, to speak for the people with our country's government; to represent them to public officials. When I grow too old to lead effectively, I will sit with the old men of the village and choose a young man with good abilities as the new chief. Then I will teach him and stay at his side to advise. This is what happened to me."

As it turns out, the Hadzabe also have pretty strict beliefs, which they claim to have held tenaciously throughout the rise of Modern Western society in Tanzania. Still, most of these beliefs boil down to one concept: living purely inside nature—existing with the planet instead of pushing it to conform to your needs. No manufacturing. No farming or husbandry.

"It is the way we have always lived," Julius says. "For thousands and thousands of years, this life has worked for us. We have gone forward in time, generations on generations, working with the Earth that provides for us. So we stay with these beliefs. In some ways, we must accommodate the outside world as it comes closer and closer to our home. But in other ways, we choose to leave the outside world alone. And this is fine, too."

A few years ago, Julius says, the Tanzanian government wanted to make life slightly easier for the Hadzabe and ordered several water wells and cisterns to be built on Hadzabe lands, giving the tribe guaranteed sources of water in times of drought. "We don't use the wells and water in the tanks, though," Julius adds. "Not that we are not grateful for the generosity. Still, we prefer to take our water as it moves over the ground. We want our water to come directly out of a river or a spring. Our belief is, natural is better. It simply is. So if the water has gone through metal, we prefer not to use it."

It's the same with housing. About the same time the government ordered the wells built, it also had a boarding school erected at the edge of the Hadzabe's valley, so the tribe's children could locate in one place and learn. "While a few Hadzabe parents send their children to the school, so they can be educated and have more options," Julius says, "most parents do not. They don't want their children sleeping under metal roofs. Again, from the way we see the world, this is not healthy. It is not our way."

Consequently, the school is now filled with the children of the outlying herding tribes, the Masai and the Datoga, who arrived with their livestock a few years ago, as their own traditional lands had been heavily grazed over. In the last few years, to help rein in grazing, the government has stepped in and further protected the Hadzabe lands inside something called the Ngorongoro Conservation Area. "Now the pastoralists can graze here by daylight, but they can't live here," Julius says. "They must leave every day at dusk."

Do the Hadzabe ever eat a goat or cow left behind?

"Never," Julius says, a moue of distaste crossing his face. "First of all, any grazing animal is not ours. So that would be wrong. And a goat or a cow is not wild. It may even have been treated with injections and medicines, so it is unfit for us to eat. And anyway, we have enough to eat. You'll see."

Julius gives a small, confident, welcoming smile, then he suggests that it is time for him to head back to his camp and get some rest. I don't doubt that what Julius has told me is true. Over the last hour, he has been welcoming, honest, and has even granted me, a visitor, my own space on his tribal lands. I like him already.

Over the next week, as I live, hunt, and gather with the Hadzabe, it is shocking how much food exists inside their home landscape. And I'm not talking simply crunchy, hastily seared fresh warthog ribs and shoulders, but really, truly, satisfyingly tasty nutrition. There are tubers and fruits and berries and grainy flatbreads. The diet is varied, and it exists all around them. It's as if they live in a grocery store.

Because my arrival comes a few months after the rainy season, which this year has been richly ample, most Hadzabe days start with a snack of fruits and berries. There is honey, gathered from inside the tops of hollow baobab trees. And seemingly everywhere beneath the ground lie clutches of enormous tubers that can be dug up with sharp sticks. Steamed in a fire for an hour or so, these tubers are then eaten with the hands, and the taste is not unlike what we Americans savor when we go after a baked potato at a Western Sizzlin.

Part of this mastery of the landscape, certainly, is that the Hadzabe have lived here for thousands of years, so they knew just where to look for different kinds of food. But with each passing day, I eat with them and never go hungry. The menu is generally two big meals daily, a late morning repast and a dinner a few hours before sunset, plus a daylong diet of small snacks. As we live and eat together, the balance of life inside their world becomes more and more obvious.

Day to day, beyond the dietary staples of meat and tubers and berries, the center of the Hadzabe diet is a dry paste made from the fruit of the baobab tree. In hunter-gatherer Africa, baobab fruit hangs around the daily diet the way high-fructose corn syrup does at today's American convenience store: it goes into everything. Unlike the high-fructose corn syrup going into soft drinks and candy bars and Twinkies, however, the fruit of the baobab is powerfully good for the human organism. When it is crushed between flat stones and mixed with water and a little berry juice, baobab fruit becomes a thick, tangy, and nutritious paste that's astonishingly rich in antioxidants, potassium, and phosphorus. It contains six times as much vitamin C as a similarly sized orange and twice the calcium of a standard, eight-ounce glass of pasteurized milk. Taken as a whole, the Hadzabe menu of charred meat and berry-flavored baobab fruit paste may sound a bit monotonous, but food doesn't get any more natural and healthful. A "flavor mimic," baobab fruit also takes on the taste of whatever it's served with. And it's also just filling enough to leave you satisfied without feeling weighted down and bloated.

This from my notebook for June 23: *Back in camp, the women crush the berries and add their juice to the crushed and reconstituted (w/water) baobab fruit. The mixture is like a thick, doughy oatmeal with fruit added. Everyone eats a lot, and when they begin to run out, the women add more crushed baobab fruit and water, until there is plenty again. The amount of food here seems virtually endless. . . .*

When I suggest this to Julius, he smiles. "I told you," he says. "We eat everything. Well, this is not true. We don't eat the hyena, because the flavor is not good . . . but it is more than that. When someone dies, we bury them and the hyenas dig them up and eat them, no matter how many stones we pile on top of the grave. The hyenas are relentless, and we have a difficult time eating something that has eaten a human. Still, when necessary, it has been done. And hyenas taste bad. But we eat pretty much everything else."

While the Hadzabe may remain isolated by both Tanzanian federal land protections and the tribe's own slightly reclusive beliefs, their way of life is beginning to reemerge as a example of healthier living to groups who've long ago left hunting and gathering in the past.

The baobab fruit's nutritional value, for instance, is becoming so well recognized that starting in 2008 it was approved for sale across European markets, with the U.S. Food and Drug Administration also in the final stages of testing it for sale in the American market. The dry pulp will be sold as an additive ingredient for energy bars, high-nutrition peanut butter and jams, and action-packed yet tasty fruit smoothies. It's strange to think that the simplest of human diets, the nutrition of hunter-gatherers, is becoming a building block in the new, science-based, Western model of nutrition, right alongside the modern world's genetically engineered corn and milk.

But as the science of making food more nutritious moves forward, the science of assessing what foods promote good heath is also advancing. So it turns out that, to make Western diets healthier, more modern nations are finally catching up with many of the ancient ways they left behind thousands of years ago in the valleys and plains of the hunter-gatherers.

As time passes, more studies are showing that, even today in the right circumstances, hunting and gathering can promote long life with limited effort. Among the Hadzabe and San and !Kung Bushmen, for instance, hunting bands remain small and, as they're family based, they integrate efforts and work well together. And their subsistence culture requires only a limited effort to provide for most material needs: three half-days of hunting and slightly longer periods of gathering is all that's often required to feed the group across any given week. Ironically, despite the perceived difficulty of hunting and gathering in a more modern world, evidence also suggests that hunter-gatherer bands have high percentages of older people, and that hunter-gather life expectancy is in rough parity with normal Western life span boundaries. In addition, the hunter-gatherers show remarkable resilience to hazards like illness, climatic change, migrations and extinctions of food sources, and accidents.

We could all benefit from a little hunter-gatherer behavior in our lives. "No matter who you are and where you live, good health comes down to the same thing: the right combination of diet and exercise," says Linda Van Horn, Ph.D, a professor of preventive medicine at Northwestern University's School of Medicine. "In order to be really, long-term healthy, every individual's diet and exercise need to complement one another. And among those people [hunter-gatherers] it has to, or they wouldn't have survived for so long."

Van Horn, who is also the editor of the *Journal of the American Dietetic Association,* has studied hunters and herdsmen around the world. "For herding or hunter-gathering groups, they need a diet that leaves them lean, strong, and fast. For them, unlike us with our houses and cars and convenience stores, diet is a component of their survival. Ours isn't anymore."

And consider this: According to Van Horn, the average American intake of high-fructose corn syrup, now the base for everything from soft drinks to virtually every snack food, has increased 800 percent in the last five years. Or this: On an average day, every day, the median American eats the fat equivalent of a stick of butter. "In our culture, we're no longer eating the kind of diet that supports our lifestyle," Van Horn says. "That's why statistics say that for the first time in many, many generations, today's young people are born with shorter life expectancies than their parents. Despite the wealth and medical expertise we have, our diet is destroying our health."

Luckily, says Van Horn, the problem is fixable, because much of the world's growing weight problem isn't simply about diet but also about factoring exercise back into our daily energy equation. "A balanced mix of healthy diet plus exercise is central to our health and longevity," she says. "African hunter-gatherers or Mongolian herdsmen are on their feet all day, chasing livestock and moving around. As a culture, we're much more sedentary. We're the ones who need to get back in balance. And it's not that hard to do. You don't have to run marathons, you just have to go out and elevate your heart rate. It doesn't have to be intense, it just has to happen."

It turns out that Dr. Van Horn understands the Hadzabe lifestyle. Each day, Julius and I and members of the clan walk out into the savanna in search of food. When we have finished hunting, gathering, and eating, securing enough nutrition for the rest of the day, the women make beaded jewelry or grind more baobab fruit between stones or gather herbs for traditional medicines. The men sit and talk . . . and talk . . . and talk. Always, though, they are doing something as they talk and share, exhibiting what anthropologists call "kinship," where the men with status within their small band maintain the floor—Julius and the elders speak most—with the younger men chiming in when conversation slows. And all the while they are sharpening arrowheads or fashioning new arrows or spear shafts from straight stalks of springy cane. Because of their central position throughout the Hadzabe hunting and diet practices, bows, arrows, and spears occupy a huge position of respect among the Hadzabe men. Throughout the day, using a blade supported vertically between their sandaled feet, they strip long, thin lengths of material from each cane's outer skin, laying it in a pile. Eventually, these strips are woven into baskets or plaited into lengths of cord. And all the while, the children play: sometimes with the dogs, sometimes in spirited games of tag or a sort of "I've got the flag, try to catch me and take it" group hunting exercise that strays far out into the savanna. Sometimes, they practice their archery skills. Along the way, during slow periods of work, both adult sexes do a little household cleaning. On one afternoon late in the course of my visit, they created a new house by bending 12- or 15-foot-long sticks to form a rough shelter armature, which they then covered with thatch. Mostly such structures are for shade and privacy, plus warming protection from the evening's sometimes cool, high-altitude breezes. The larger point is: There is always something to do, but rarely does this agenda carry any scheduled pressure with it. Things happen as they happen.

Are the Hadzabe ever bored? I ask Julius.

"There is always something to do," he says.

---

According to the anthropologist Pauline "Polly" Wiessner, Ph.D, of the University of Utah, the daylong industry of African hunter-gatherers like the Hadzabe serves several different purposes. "Among African hunter-gatherers, all day long, there's constant social interaction," Wiessner says. "It's part of their social structure, and is a form of their social control."

Wiessner has spent decades studying click-talking !Kung-language hunter-gatherers, who came into existence around the same time as the Hadzabe, and live southwest of the Hadzabe in Namibia. She learned to speak the language during her research. For three decades, she has returned to live with these people for long periods of fieldwork, often stretching to several months.

"Because of their limited resources, African hunter-gatherers are always watching, always monitoring what goes on in the group," she says. "It's a question of survival to them. Who is sharing with whom? Who is withholding from whom? Really, a lot of it is just like our workplaces and neighborhoods. It's gossip. It's office politics. It's social control. But it's also the spice of life, you know? It's what people do."

During one of her three- or four-month field seasons with the !Kung speakers, Wiessner has done regular breakdowns showing what each day's collective conversations, which percolate from dawn until well after dark, were composed of. Among other things, she learned that daytime talk among hunter-gatherers is often rooted in need and directed toward using resources, whereas nighttime discussions are "far more devoted to storytelling and collective understanding. There's a real difference in the content between the two."

This was funny, and it says volumes about the value of language inside a society. When I sat with the Hadzabe during my visit, listening to them talk as they worked away on different projects, I was oblivious of any social control, gossip, or political content. What I felt was total, relaxed, happy community.

"That's because you don't speak the language," Wiessner says. "If you did, you'd find they're constantly monitoring social interactions within the group. You have to understand, the technology that first carried modern humans away from the lives of their predecessors was at its basis an understanding of simple social cues. Among our ancestors, increased types of communication were the first real leap forward in technology. Social networking is what initially launched us. It wasn't technology in the way we think about it today. It was a solid social matrix. It allowed for everything that came after. And we have to remember that."

During my week with the Hadzabe, in fact, when we are not hunting or "working" on daily chores, the huge periods of free time are filled with experimentation inside the world around them.

Each time they find something new—my small, battery-powered Petzl headlamp for example—they examine it with almost scientific closeness. Once they understand the headlamp's externals, they dismantle it; this is followed by testing its elastic strap and feeling the heft of its AAA batteries. Showing a remarkable sophistication for a people wearing animal skins and sandals made from old car tires, they lay out the headlamp's constituent parts, wondering what else they can use these ingredients for and talking among themselves. Finally, when they have exhausted the afternoon's possibilities with the headlamp, they put it precisely and expertly back together.

Wiessner agrees with this characterization. "When I'm repairing a car, I always make sure to have a couple of the !Kung watching, since when it comes to mechanical things, they remember everything. They're amazing. But, then, part of that speaks to the fact that they don't have all of the things, the material resources, that we have. Here in the Western world, in our daily lives, we couldn't know how to take everything apart, to see how everything works. We use too many things every day. We wouldn't have the time . . . and all that complexity would make us crazy. But if you're an African hunter-gatherer, your available resources are far more limited. . . . They just don't have that many complex objects to deal with, so they're able to memorize those complex objects they do come into contact with far more easily."

Whatever the reason, every day I am with the Hadzabe, I love watching how their minds work: how they come to know new things. In fact, over the ten or so days I'm with the Hadzabe, I develop an almost jealous respect for what they are capable of doing and the openness with which they approach the world. Though isolated, they are remarkable and wonderfully resourceful and intelligent people. In some ways, in fact, they remain willfully isolated, not buying into the ease of a more modern, more Westernized life that encroaches all around them. Though they don't see outsiders much, they are not threatened by visitors possessing Toyota Land Cruisers and high-tech tents and headlamps. Instead, they use what they want from this other

*Another day as it's always been: the Hadzabe of Tanzania sharing food and warmth and stories around a fire*

world, and disinterestedly leave the rest, happy in the lives they are living. Most nights, around sunset, they make a nice, warming fire and then settle in for more talk.

Every day among the Hadzabe, raw human curiosity and ingenuity are on display; traits I rarely see publicly in the Western world anymore. Sure, in our world computer programmers, video-game designers, credit-default-swap financiers, and chefs and artists invent towering creative works out of thin air. But with the Hadzabe, experimentation doesn't seem to be work directed at a defined end point, like a new off-Broadway musical or a public stock offering or the newest Nintendo Wii iteration of "Super Mario Brothers." More than anything, the Hadzabe experiment with—and in—the world around them. It isn't work. What they are doing far more resembles play.

One evening, my belly full of a dessert of baobab paste and the crushed juice left by husks of piquant and tiny black berries we'd collected earlier in the day, I can't help feeling extremely fulfilled and at peace. The days are characterized by hunting for meat, gathering berries and tubers, and the simple and pleasant satisfaction that comes from cause-and-effect effort. In the twilight, as the first stars begin peeking from the blue scrim of sky that follows sunset, I sit in the dust, my back against a tree. A couple of the band's men are reviving fires from embers that have smoldered all day. Life feels exceedingly pleasant, and is moving at a pace that's both comfortable and completely comprehensible. The experience is placid in a way I rarely come to find at home.

Though I am still new to the ways of a hunter-gatherer existence, its rhythms feel completely natural, as if I'd been doing it all my life.

Had I been home in Virginia, even if I'd just eaten a meal out of our family's garden, I'd still have been wondering about what we were going to do next, what we were going to eat tomorrow and the next day . . . and five months from now, when the garden was finally exhausted, its force knocked flat by November's frosts. The TV or radio would be on. There would be homework to double-check, bills to consider, that nagging worry about my pickup truck's faltering water pump in the back of my mind, its failure waiting to step out like a menace from behind a tree.

For the Hadzabe, living simply and almost serenely in their own grocery store, I wonder if the details of their day-to-day living ever get them worrying.

"Do the Hadzabe ever run out of food?" I ask Julius.

"No, there's usually enough, there is usually more than enough," Julius says. "But, yes, eventually, we do hunt and gather an area down to lower supplies. When food day-to-day becomes more difficult to find, the elders and I have a meeting. We send out five boys to look for new land. They go for two or three days in different directions, and then come back with reports. After we hear their observations, as a tribe, we decide where to go next on our lands. Sometimes we travel as far as 60 kilometers. It doesn't happen all that often, though. We move every six to nine months. Usually we move to a place we have been before, historically. So we know what is there and what it looks like. We might even move back to a settlement site we've occupied before, though that is not always the case."

In the near darkness, Julius pauses for a long time. He nods, and a small smile crosses his face. "Sometimes," he says, "it's fun to start somewhere new. There is adventure in starting somewhere new, learning about it. And that kind of living is good and healthy. It is good to keep some challenge in your life and not get too comfortable. We do well with challenges, and this fact is as true today as it was when our common ancestors lived here together so long ago."

Among the Hadzabe, every hour brings another lesson: a revelation.

The afternoon following the warthog, as we sit and talk in the shade, dozing and joking, Julius tells me about his 2005 trip to Washington, D.C., for the announcement of the

Genographic Project. “I was happy for the new experience, but I was not in my usual place,” he says. “That made me nervous. On the airplane, I didn’t feel very well. I felt confined and the pressure on my body made me sick. On the airplane, I thought I was going to die. But when I got there, with the different food and all of the people and the crowds, it was interesting. I got a bed. It was very comfortable. The people were very welcoming to me. But it was cold. I was cold all of the time. I was very happy when I got to return home. I am happiest here, in Tanzania. Still, making a visit to Washington, D.C., was something to do once.”

The time passes. As the sun moves lower, Julius announces that it’s time for me to “really understand what we are talking about when we talk about being related as family.” He stands, gestures for me to stand, too, and then points into the savanna, implying that we’ll be taking a walk. Then he heads off into the brush. I follow. After 15 minutes of walking, we’re approaching an enormous baobab tree: big around as a house and plugged into the dirt in the midst of a broad plain. To the northwest, the cliffs of the Rift Mountains are so brightly lit by the sun that their rocky features show clearly. They appear closer than ever.

On a branch high in the baobab, perhaps 80 feet off the ground, a cloud of bees swarms against the sky. It’s clear they are from a large hive, living inside the branch. The honey inside makes the tree’s bark slick and shiny. “See,” Julius says. He points at the branch. “That’s good. The honey is new and fresh and almost ready for us to collect. We will harvest that honey next month. Usually, most years, the bees in that branch give us about 80 liters of honey, which is enough to keep us for a long period. Honey is a favorite of ours. What we don't use, we take to town. There we trade it for things we can use to make into arrow points, or we trade it for pots and pans for cooking. Sometimes the men in my group trade it for tobacco or for a little marijuana to smoke in the night."

Along the tree's trunk, Julius points to a set of steps: two-inch-diameter sticks that have been jammed into holes bored in the tree like the rungs of a one-sided ladder. The steps stretch from the ground all the way to the hive. "But showing you one of the places we harvest honey is not why I brought you to this tree right now," Julius says. "Instead, this is what I wanted to show you."

We walk to another side of the baobab's gigantic trunk, where a large, cave-like shelter has been carved into it: an excavation big enough to sleep several people. "This is where we Hadzabe have our babies," Julius says.

I stick my head inside. The hollowed area inside is tall, extending maybe 20 feet upward, like a chimney. There is a small hole at the top, which lets in sunshine, illuminating bats that flap chaotically near the opening, disturbed by my arrival. Seeing the bats, I pull my head back out.

"Inside this tree, the expectant mother is joined by her mothers and sisters, and they often spend a month or two months here," Julius continues. "This is where we Hadzabe have been born for 1,000 years . . . or maybe many thousands of years more than that. This is where grandfathers of grandfathers of grandfathers have been born. This is where I was born. I really,

really like it here. Every time, I find peace here. Sometimes, I come here and sleep on the ground, just because of the sense it gives me. For us, this tree is life. Only when a mother knows her baby has become strong enough to leave this tree will she return to the village. For us, this tree is the world. We are all descendants of this tree."

"Am I a descendant of this tree?" I ask, wondering aloud. Its distended and strange limbs remind me of Luca Cavalli-Sforza's deep genetic diagram.

Julius shrugs. If it is true that, on a deep, ancient and genetic level, he and I are really family, then for him the answer is obvious. He looks me in the eye, and gives another of his friendly, confident, quick smiles. With a characteristic flicking gesture, he tosses an open hand at the baobab. "We are all descendants of this tree," he says.

Then, as if to conclude, he smiles again. "Come on, let's go back to the houses," he says. "There will be food soon."

After that, with the late afternoon sun falling across the landscape and dousing everything from the mountains to the trees and grasses and people in an even richer, deeper gold, Julius and I walk back toward his camp—and my camp beyond.

As we go, Julius begins pointing out plants whose roots, leaves, and nuts are thought to have medicinal qualities. He stops and touches the leaves of a small shrub. "This one is called *mokuei* or *kanni quea* . . . we crush it and use it to treat coughs," he says. "We make a kind of tea with it."

Another bush, just a few steps away, is smaller, with greener, more elongated leaves. "This is *madazgue,*" Julius says. "If a mother is breast-feeding, especially a new daughter, you crush this up and the mother washes with this. It helps to create milk for the baby."

We walk another five or ten steps, then Julius grabs a spindly, almost dead-looking shrub that rises past his shoulders. "This is *undepicala,*" he says. "Also called *congo lok.* This one is a very special plant for us. It gives life in many ways. Its stems make very good arrow shafts, because the stems grow very straight. The roots also provide snakebite treatment. One of the men in my family, the small and very black one called Ma Yoyo, was bitten by a puff adder last year. You can still see how his leg is swollen. It was the roots of this plant that saved his life."

We keep walking. Julius points out another shrub he calls *mugalombea.* "If you want to make a daughter with your wife, you cook the leaves of this plant in a soup. It is very strong, both in flavor and effect. You make the soup with this, and you get a daughter."

As we continue moving back toward camp, Julius leads the way and keeps pointing out plants as we go. The leaves of one small shrub calm upset stomachs; another is applied directly to the skin before honey-gathering to dull bee stings; another cures fevers when ground up and mixed with hot water. At one medium-size plant, whose leaves look desiccated and dry, Julius pauses. "These leaves are where the poison for the arrows comes from. You grind the leaves down, and boil them in water until it makes a paste, then you rub that paste on the arrows. It works."

"And all of these work reliably?" I ask. "These all work for you?"

"Oh, oh yes," Julius says. "We have hosted Western doctors and researchers who have come here before you. And, yes, they were surprised, too. But for us, we have always lived here. Our long time living here helped our ancestors learn these things, and that knowledge has been passed along to us, and we pass it along to our children. So why wouldn't we know these things? This is our home. You know the plants and animals around your own home. Isn't this so?"

"Yes," I say. "To a degree. But not like this."

"How long have you lived in your home?"

"About 15 years," I say.

Julius stops. In the gilded light of sunset, he turns and stares back at me, his eyes dark and squinting slightly; a confused smile settling on his face. "We live in the same world . . . " he holds up his right hand, palm upward, "and we live in different worlds," he continues, lifting his left hand in the same gesture. Then he moves his hands up and down, as if juggling invisible blocks from hand to hand.

After a second or two, with another of his quick hand movements, Julius flicks his right hand over his shoulder. "We go," he says. And turning to face his camp again, he begins walking home.

*A young Hadzabe woman contemplates the world from beside the baobab tree where all the Hadzabe are born.*

# WORLD OF POSSIBILITIES

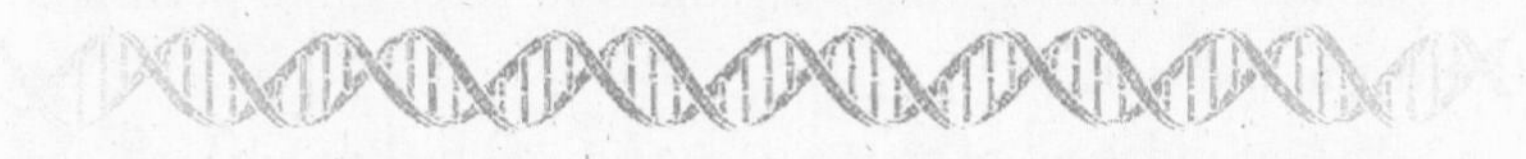

**BASICALLY, THE RISE** of all life on Earth to where it exists today—everything from dinosaur fossils to chemical sprays that kill broadleaf weeds; endangered rhinos and ever faster computer chips—must eventually be boiled down to just one of two possible explanations. And for anyone who actually thinks about it, you're forced to pick a side.

The towering irony implied by this choice (or maybe its flabbergasting poetry, depending on how you see it) is that both sides leave as much unanswered as they explain.

The first and older choice, and one that some scientists and rationalists find incomprehensible, is that of Divine Creation. If a person carries religious leanings or beliefs, no matter what the religion, there is a creation myth at their faith's center that each believer must address and integrate into his or her views. Viewed from this side of the ledger, everything was created by some force greater than the combined content of the universe itself. And when you ponder it—mass and light and space and time . . . and the improbable rise of life—that's a pretty tall order. Still, for those people with a well-integrated creation myth, it all makes sense and—this is important—provides them with a modicum of framing, allowing them to erect some peace and order in the world.

The second view, of course, is the rise of life through early cosmological and, later, biological accident and evolution, as studied, charted, and somewhat explained by science. Still this view leaves so much still unexplained near its foundations that it, too, should be accepted as a form of secular religion. The universe, it is said by many who follow this set of beliefs, was created at least ten billion years ago in an instant: by the so-called big bang resulting from a violent and enormous reaction inside a dense cloud of heat and pressure.

Yet this scenario contains its own mystery. Where, uh, did the dense cloud of heat and pressure come from?

It's not inconceivable to think that science may one day explain everything in the universe. But even with the great leaps forward in technology and understanding that have been achieved over the past century, few reasonable people are holding their breath that the universal unifying theory, as it's sometimes called, will be unveiled any time soon.

There's a rich paradox (perhaps enjoyed by me alone) in noting that both sides of the discussion must eventually rest their beliefs on different articles of faith, and that neither can readily accept the other's version of how events in the deep past have gone. And as much as it pains rationalists to admit: In science, just as in religion, numerous core mysteries prove elusive.

Still, in the last few centuries, science has been generating more and more plausible and provable explanations for how life on Earth arose from what was originally a piece of rock adrift in space. And while other books can chart for you the rise of all life on Earth (thanks, David Attenborough or Bill Bryson!), for our purposes the story starts a little under two million years ago, with the evolutionary arrival of *Homo erectus,* the "upright man."

More robust but structurally similar from the neck down to modern humans, but with a far smaller relative brain capacity, *Homo erectus* appears to have been, 1.8 million years ago, the latest flower following the genetic split from the apes that occurred sometime more than five million years ago and likely led to our first true human ancestor. And the human race appears to have arisen on the plains of Africa. The archaeological record shows that *Homo erectus* employed higher thought in the process of making primitive stone tools and, possibly, even using rudimentary gestures to convey ideas.

And *Homo erectus* hung in there for quite a long time, evolutionarily speaking. Climatically, the world across that time was, away from our forebears' equatorial African territory, often cold and unforgiving. According to the geological record, ice ages washed across it at intervals. But during this time, in Africa, *Homo erectus* appears to have thrived, prospered, and even spread widely. While most everyone who watches these things is in agreement that the species originated in Africa, *Homo erectus* eventually migrated to eastern parts of Europe and other parts of the globe by at least one million years ago. Fossilized *Homo erectus* remains have been found in Africa, Georgia, Indonesia, Spain, Vietnam, and China.

"I worry, though, that the reason we've called all of these specimens from all of these places *Homo erectus* is because we have so few examples of it, even if those specimens are widely distributed," says Alison Brooks, a professor of paleoanthropology at George Washington University and a widely acknowledged authority on early humans. "I think that the more we learn about early hominids through fossil evidence, the more we're going to realize that the history is far more layered and complicated than we want to believe. That's life. Eventually, I think we'll come to understand how these species evolved and were distributed. But to lump them all together, from all these different localities in these various places, and to say they're all the same? Over all those thousands or millions of years and that kind of geographic distribution? Well, it's almost got to be more complicated than that."

Some scientists have taken this reasoned approach and run in an entirely different direction. Over the last century, a small group of scientists and explorers, such as Debbie Martyr of Britain's Fauna and Flora International most recently, believe *Homo erectus* may still walk the earth. Citing some discoveries in the more recent fossil and natural history record, they believe the species may continue to reside in the isolated jungles of the Indonesian island of Sumatra. In 2004, in the scientific journal *Nature,* for instance, a description of what's become known as *Homo floresiensis* (a possible *Homo erectus* relative) was published. A small-statured, three-foot-tall species of hominid biped found in a cave on the Indonesian island of Flores a few years earlier, the specimen's bones, plus some small stone tools, were dated to about 12,000 years ago, meaning it would have coexisted in the region with modern humans for something like 25,000 years.

But in the world of *Homo erectus* in Indonesia, science and anecdote continue to spin even larger conspiracies. Known there as *orang pendek* (short people) by the locals, these beings are allegedly seen with regularity by jungle-dwelling natives and sought after by Western scientists. In fact, in 2006 and 2008, the National Geographic Society sent a scientific team into the Sumatran jungles with motion-detecting camera traps in hopes of snapping a photo of such a creature.

While I heard often of *orang pendek* when visiting small jungle villages in Indonesia in 1996, 2000, and 2005, there were no supporting data beyond the anecdote, and the fossil and archaeological description in the *Nature* article from 2004.

There do exist several credible, if aging, Western anecdotes of eyewitness evidence to support the "*Homo erectus* still lives" theory. Like this one: In 1923, a Dutch explorer and hunter named Van Heewarden described his encounter with it in the Sumatran jungle this way:

> I discovered a dark and hairy creature on a branch. . . . The sedapa was also hairy on the front of its body; the colour there was a little lighter than on the back. The very dark hair on its head fell to just below the

> shoulder-blades or even almost to the waist. . . . Had it been standing, its arms would have reached to a little above its knees; they were therefore long, but its legs seemed to me rather short. I did not see its feet, but I did see some toes which were shaped in a very normal manner. . . . There was nothing repulsive or ugly about its face, nor was it at all apelike.

Yet with a full-grown stature said to be about half the height of today's modern humans, and a skull far smaller proportionally than ours, many anthropologists note that while aspects of today's natural world still remain unknown, if *orang pendek* or *Homo floresiensis* exist at all, they are likely a band of shy bipeds that somehow remain capable of surviving among the most demanding and impenetrable jungle environments on Earth.

Still, whether you subscribe to the idea that tiny human cousins are still living in Indonesia or not, older aspects of the fossil record show that *Homo erectus* succeeded over the first two million years of its existence, with its skills and social abilities verifiable through archaeological evidence.

At the Zhoukoudian clefts near Beijing, China, a famous *Homo erectus* fossil locality, home to what was called Peking Man before it was recategorized into *erectus* in the 1960s, controversial carbon-dated evidence in the form of animal bones and presumed fire pits tentatively shows that 400,000 years ago,

*Homo erectus* was hunting and killing large animals (including camel, wild boar, deer, and rhinoceros), and then butchering and cooking the prey in fire pits inside the clefts.

Judging by the findings surrounding Peking Man, *Homo erectus* could hunt and wield fire, and had migrated away from Africa across great distances over thousands, or tens of thousands, of years. And while this evidence points toward *Homo erectus* being one of the first upright bipeds to use higher thinking and fire and simple stone knives and Acheulean hand axes to make itself more comfortable in the world, its demise as a species has never been explained.

In this world of human possibilities, despite all that science has been able to uncover about the history of humans and the Earth, central mysteries do remain.

With regard to *Homo erectus,* in fact, its very existence and broad geographic distribution are about the last facts all anthropological and scientific sides agree on. After that, with the subsequent rise of Neanderthals and the modern human, *Homo sapiens sapiens* (or Cro-Magnons, as the early European ancestors of our race are also sometimes called), few scientists seem in harmony about what precisely happened next.

"It's telling, I think, that the closer we get to today's modern humans, the nastier the disagreements [between scientists] get," says Alison Brooks. "But then, once again, that's because our knowledge of these things is still not complete. There are gaps in time and in our knowledge. There are gaps in our understanding of the fossil record. But there are some things we're starting to know with certainty."

One point of agreement is that, following the geographic distribution of *Homo erectus,* the Neanderthals (or *Homo sapiens neanderthalensis* or *Homo neanderthalensis,* depending on who you ask) came next. Experts don't agree on whether the Neanderthals represent a separate species or a subspecies of *Homo sapiens,* and that belief quickly separates them into opposing schools of thought. That said, it is agreed that Neanderthals were upright bipeds that likely appeared between 130,000 and 230,000 years ago, with their arrival date less certain than the date of their extinction, known to be about 28,000 years ago.

And from there, general scientific agreement about Neanderthals and modern humans becomes a scientific pie fight. One group of paleoarchaeologists, labeled the "out-of-Africa" thinkers, speculates that Neanderthals migrated into Europe from the African continent about 130,000 years ago, possibly through Spain, where some of the earliest (and latest) artifacts, including those stone knives and bi-face hand axes, are found. Another group of scholars, the "multiregionalists," hypothesizes that Neanderthals were actually *Homo erectus* descendants that evolved and became more robust through "hybrid vigor" and "gene diffusion" (or procreating with strangers as we call it today) across different zones around the world, with the Neanderthals of Europe being the most successful of this new subspecies.

Scientists on both sides of this argument have reasonable data and aspects of the fossil record to back up their conclusions. And, like competing industries, both publish their new findings each year by the pound, also advancing their theories at regularly scheduled and well-attended symposia.

Still a third school, much less common, believes that Neanderthals and Cro-Magnons mated, birthing hybrid humans, with the proof being "morphological," as they note that some modern humans have Neanderthal-like physical attributes, such as heavier brow lines and sturdier general bone structures. About this theory, Alison Brooks, Spencer Wells, and others are quick to counter.

"There is no genetic evidence to support any interbreeding," says Wells. "We've checked. We've done the science. And there's no DNA evidence that the two species ever interbred . . . at all."

Like I said: a pie fight.

First discovered in 1856 by quarry workers in the Neander Valley—(Neander Tal in German), near the city of Düsseldorf, the fossil remains of Neanderthals are larger and more physically sturdy than are those of *Homo erectus*. As more Neanderthal specimens have been discovered across a distribution area that spreads from Spain and Great Britain to the Middle East and Central Asia, the species, with its six-foot stature, heavy bones, more prominent ridges along the brow line, and larger average brain capacity than modern humans, was originally hailed as the "missing link" between *Homo sapiens* and the apes.

In fact, because of their heavy physical features, since Neanderthals were first discovered and identified, they have been the butt of gentle paleo-prejudice from their *Homo sapiens* cousins.

Sometimes referred to as cave men and ape men, they are regularly characterized as mouth-breathing knuckle-draggers. Even today among paleoanthropological scholars, thanks to the Neanderthals' broad noses, wide-set eyes, and heavy cranial and skeletal features, even some of the world's most acclaimed scientists privately refer to Neanderthals as "big faces."

Their lives were hard—and apparently often cold. Modern climate studies looking into times of Neanderthal habitation show that during their time on Earth, the planet was regularly beset by ice sheets and eras of low food and animal availability. Unlike the relatively stable planetary environmental conditions of the last 10,000 years, the Earth of the Neanderthal age was far more demanding. According to the journal *Science,* 17 ice ages plus several periods of glaciation have washed across the planet in the past two million years. Some of those conditions occurred during the time of the Neanderthals, while at the same time our direct relatives, *Homo sapiens sapiens,* were still evolving and sunning themselves across the warm, rocky plains of the African Rift. (A note: For purposes of brevity, from here on, I'll refer to the two species as Neanderthals and *Homo sapiens*).

Other factors during the age of the Neanderthals likely made life hard as well. Take, for example, the eruption of what today is called Mount Toba, on the Indonesian island of Sumatra, which blew up volcanically about 74,000 years ago. According to the climate scientist Michael Rampino at New York University and acclaimed Icelandic vulcanologist Haraldur Sigurdsson, a Mount Toba volcanic blast "occurred during a relatively rapid interglacial-glacial cooling and sea-level drop, which may have played a role in triggering the eruption. Stratospheric dust and aerosols could have accelerated the cooling by causing a 'volcanic winter' with global temperature decreases of 3° to 5°C and minus 10°C during the growing seasons. . . . Ice core studies indicate the Toba aerosols remained in the stratosphere for the unusually long time of [approximately] 6 years."

In contemporary terms of reference, the Mount Toba event was 100 times more powerful than the Mount St. Helens eruption of 1980, when some 800 cubic kilometers of ash filled the Earth's skies, reflecting away sunshine, cooling the planet, and filling the air with dust. Worse than Mount St. Helens, but still in modern memory, is Indonesia's Mount Tambora eruption of 1815. A mountain in close proximity to Mount Toba, Tambora is said to have given off so much ash and sulfur dioxide that the following year, 1816, is recalled across the planet as "the year without summer," with crop-killing frosts occurring in every month of the year in the temperate latitudes of the United States and Europe. Harvests were a disaster.

It's these facts that buttress arguments made by Stanley Ambrose, Ph.D., of the University of Illinois, and others, that in the wake of the Mount Toba eruption, so much reflective ash was piled up during an already established period of global cooling that Neanderthal populations were driven through a "population bottleneck" and nearly to the brink of extinction, leaving only a few thousand families to survive.

Still, hard as the existence may have been, Neanderthal life was apparently not without an appreciation of beauty and poignancy attached. One collection of Neanderthal skeletons, estimated to be between 40,000 and 80,000 years old and discovered in a cave called Shanidar in the Kurdistan province of northern Iraq, proved to be ten individuals, men, women, and infants alike, all buried by others in a common location. Dr. Erik Trinkaus of Washington University in St. Louis, who discovered and studied the Shanidar specimens, suspects they died of starvation in the late winter, when cold had been a constant grinding force depleting stored body fat, and at a time when environmental stresses were greatest and food sources at their lowest.

For Trinkaus and his colleagues and followers, what's most touching about the Shanidar burial specimens isn't that they were interred together. Instead it's this: Excavated along with one of the Neanderthal remains were samples of pollen from early spring flowers that were located near the body, suggesting that those who buried him sent him off encircled by the first signs of earth's annual renewal.

This is not the behavior of knuckle-draggers.

At Shanidar, Trinkaus and his colleagues made other discoveries, too. Take, for instance, Shanidar 1, as he's known: a male whose remains were found in the same cave.

"He's one of the few Neanderthals we have over 40," Trinkaus has said. Judging by assessments of dental enamel and bone structure, most Neanderthals excavated and studied thus far didn't live past about the age of 30.

Shanidar 1 shows evidence of a hard life: bone scarring left behind by breaks and healing along his entire skeleton. The orbital bones around his left eye and down his cheek, for instance, were crushed at some earlier point in his life. And while his skull had repaired itself, the recuperation was not without lasting physical alteration. To another Neanderthal coming upon him, his face would have looked strange and asymmetrical. The damage was severe enough on Shanidar 1's face that Trinkaus believes he may even have been blind in his left eye.

Evidence also shows that Shanidar 1's right arm was withered from his shoulder down to his hand. Healing lines along the arm bones show repair from complete breaks in the years before his death, and these suggest the arm was broken in at least two places. And on Shanidar 1's feet, judging by the joints, arthritis was present in one of his big toes. There was also a long-healed fracture on the outside of his foot: damage consistent with something extremely heavy having been dropped on it.

With a crushed left face, a broken and withered right arm, and arthritic feet, Shanidar 1 suggests that being a Neanderthal was hard.

How difficult was it? About a decade ago, one of Trinkaus's graduate students, Tommy Berger, performed an overall assessment of 17 different Neanderthal skeletons—which together showed 27 different traumatic skeletal injuries—and compared their damage and bone-break locations to modern skeletons. As it turned out, among the sample Neanderthal group, skeletal damage largely occurred on their upper bodies and skulls. And after crunching the numbers in relation to modern humans, Berger determined that the closest analogous injury pattern today is the battered bodies of rodeo riders: humans forcibly hurled to the earth for a living, where sometimes they are then stomped on by heavy, hooved animals.

Still, the Neanderthals survived and carried onward for more than 100,000 years. Judging from collections of Neanderthal bones and tools found in successively deeper strata of some localities and caves, it appears that for most of the time Neanderthals inhabited Europe and the Middle East they were gatherers and scavengers, with organized, tool-assisted hunting perhaps coming late to their skill set. For many modern observers, these findings suggest that the Neanderthals picked up these practices from interaction with a distant cousin.

As already mentioned, roughly 200,000 years ago, on the savanna of east Africa, that other new hominid species was evolving. Eventually it would be known as *Homo sapiens* ("wise man"), but it had yet to live up to that name.

Far more slender than the Neanderthals, but about roughly the same physical height, *Homo sapiens* were hot-weather sprinters to the Neanderthals cold-season bruisers. And by about 70,000 years ago, these modern humans were beginning to spread permanently beyond their original continental homeland, working east into the Arabian Peninsula and north across the Sinai Peninsula into today's Middle East.

"But what you see even with early *Homo sapiens* is different from what you see with Neanderthals," says Pauline Wiessner of the University of Utah. "Almost from the start, *Homo sapiens* were using sharpened projectiles. They had spears with points and atlatls: hand-held levers that held the spears, which allowed them to throw spears much farther and with more velocity. They also used clothing decoration and symbology: they used dress and decorations to make distinctions about themselves. Ask yourself: why would they do that? It's because they were socializing on a far wider level than Neanderthals, who didn't have similar decorative distinctions. For *Homo sapiens,* the beads and clothing design told strangers about them. Even 80,000 or 90,000 years ago, they were socializing with people outside their groups. And decorations on their clothes told strangers a little something about them. It was purely social. It was the first social technology."

Among paleoarchaeologists, "social technology," as it's known, seems to explain much of the fast rise of the *Homo sapiens* and the fall of the Neanderthals, as wave upon wave of newcomers washed up on traditionally Neanderthal lands.

"It's not that the Neanderthals weren't able to use tools," continues Wiessner, "but they didn't have the motivation to advance those tools' uses. Even today, for instance, we all recognize that some groups of chimps use tools for simple things, food gathering and so forth, but they never make the next leap. For some reason, across almost their entire existence, the Neanderthals simply didn't advance the use of existing technology. Then about 45,000 years ago, when the *Homo sapiens* arrived in the traditionally Neanderthal range in Europe, they outcompeted the Neanderthals because of the advances underwritten by their social networking; their social technologies."

Wiessner says that studies of youth and infant tooth enamel in Neanderthals and *Homo sapiens* shows that Neanderthal infants breast-fed for far longer.

"Why would that be?" Wiessner asks. "Because the *Homo sapiens* may have been capable of finding wider and more ready and reliable stores of food thanks to these broadened social networks. The Neanderthals lived in small groups; the *Homo sapiens* seem to have existed in far larger networks. That socializing led to the advancing of other technology, as well, but socializing was the real foundation. You have to view the social matrix between groups being, basically, not only the first technology, but the thing that also drove a lot of the physical technologies—the spear points and things—that developed later. Social interactions and watching what others were doing pushed our early technologies forward."

Remarkably, the archaeological record shows that, from roughly the arrival of *Homo sapiens* into traditionally Neanderthal ranges until the demise of the Neanderthals took only about 4,000 years, or about 750 generations.

While from today's perspective, 750 generations is a long time, in terms of the extinction of an entire intelligent race that had existed about 20 times that long, it becomes obvious that something, or perhaps several factors, inside the world of the Neanderthals changed. After all, they had proved able to deal with the demanding European weather, the volcanic eruptions that possibly set off ice ages, and the hardships of scraping life from a planet sometimes stingy with food. Because of this ability to cope with prehistoric Europe's capricious demands, many authorities believe the extinction of the Neanderthals was most likely due to the introduction of something new to their lives.

"When you think of the speed of that change, after maybe 90,000 years of existence, you have to ask, Why?" says Alison Brooks, "I mean, either it was the introduction of some pathogen into the Neanderthal society, or it was a technological difference between the two groups that left the *Homo sapiens* outcompeting the Neanderthals. In the same way a hunter with a rifle can outcompete a hunter with a bow and arrow or a spear, these new kinds of technology, plus the rise of *Homo sapiens* social technologies, just tilted the balance of survival in their direction. Some people who study this believe there was war between the two species, but they might just as well have coexisted in the same areas, with the *Homo sapiens* simply being better hunters or being able to supply more food."

According to Brooks, the archaeological record literally shows very little or no use of small projectile points by Neanderthals until roughly the time *Homo sapiens* arrived in traditional Neanderthal ranges. "So Neanderthals were capable of learning . . . of adapting technology from others," Brooks says. "But for some reason, they couldn't do it for themselves. And the record shows that *Homo sapiens* were also doing all of this symbological stuff: they were making beads and paintings in caves and even fabricating small sculptures of bone and rock. Thanks to their social technologies and wide networks of interaction over huge areas of land, their ability to thrive was only growing. And even though the Neanderthals were adapting some of these newcomers' technologies into their lives, the same can't be said for them. And that might have proved the difference between one species' thriving and another dying out."

In short, *Homo sapiens* bested the Neanderthals across Europe, not through physical strength or even familiarity with the landscape. Instead, they did it with social networks and improving technologies: the use of art as storytelling, plus symbols, and new means of acquiring food. And beneath it all, the engine for this leap forward was an underlying motivation to communicate, even with strangers. They were watching, thinking, communicating, and exploiting opportunities.

Around 34,000 years ago, in a rock shelter called Magnon in the Dordogne region of southern France, *Homo sapiens* began drawing on cave walls: pictures of bison, horses, deer, animals they were likely killing for food and wanted to commemorate. These images are still there today. At Magnon, the bouncing beams of flashlights can still illuminate those paintings: a direct link to some of our more ancient ancestors. And as a locality for early evidence of *Homo sapiens* in Europe, the Dordogne is not alone. In fact, there are many other sites in southwestern France and northern Spain with similar art left on the walls; in the interest of preserving these ancient galleries, they are being kept from public view.

And the newcomers to the region, *Homo sapiens,* weren't simply drawing on walls: they were advancing their culture in other new and remarkable ways, as well.

---

Over the past two years, working in caves in southwestern Germany, near the city of Ulm, archaeologists have been unearthing fossil evidence showing that *Homo sapiens* didn't simply paint their cave walls in pursuit of self-expression—they had a far more involved artistic tradition than previously believed.

This is most evident in the discovery of several flutes from three different caves. Each of these instruments has a carved mouthpiece and is made from the hollow wingbones of griffon vultures, and all are said to date from roughly 35,000 calendar years ago. The flutes have several finger holes, to allow the person playing them to create different sounds, and some show scoring along the bones, perhaps indicating where the fingers should be placed.

"These demonstrate the presence of a well-established musical tradition at the time when modern humans colonized Europe," says Nicholas J. Conard, Ph.D., of the University of Tübingen in Germany, who found the flutes. After what can only be assumed to be recent trials by scientists involved in the find, Conard also adds that the flutes are capable of making "tones that are quite harmonic." Dr. Conard then goes on to note that the most recent flute find was unearthed in sediments just a few feet away from a carved figurine of a busty, nude woman. It's one more scrap of anecdotal proof that the creation of art beyond music and cave painting was also ongoing in the earliest societies of *Homo sapiens* in Europe.

Underscoring the "social technologies" belief, Conard also adds that the Neanderthals were existing in caves nearby at the same time, and that they left little or no evidence of musical or artistic traditions in any of their dwellings. Consequently, Conard and others believe that examples of music and art made by *Homo sapiens* "could have contributed to the maintenance of larger social networks," Conard says. "And thereby perhaps helped facilitate the demographic and territorial expansion of modern humans."

But from cave paintings to finer handmade objects to music and primitive but representative stone sculptures, as science drills deeper into the complex history of modern humans on Earth, one thing grows clearer and clearer. As *Homo sapiens,* the great communicators and experimenters, started spreading across the world, they were taking art and music with them. And using these forms of expression and symbols, all of them signs of increasing levels of social connectedness even between relative strangers, they were starting to make huge leaps forward in social and technological progress.

But the rise of modern humans wasn't all about a small, smart, resourceful population suddenly washing out of the warm plains of Africa and across a more climatically indifferent planet. While some *Homo sapiens* hunter-gatherers migrated away from Africa in radiating waves to the Middle East, the Indian subcontinent, and the plains of Central Asia, eventually circling westward into Europe, others stayed home.

In fact, recent genetic surveys of people in Africa by Sarah A. Tishkoff of the University of Pennsylvania, in the largest survey ever of its kind, found that the click-talking San Bushmen of the Kalahari Desert, to the southwest of Julius and his Hadzabe people in Tanzania, may carry the oldest human DNA on Earth.

Which circles us all the way back to the anthropological DNA work of Spencer Wells and others, the Genographic Project, and my journey.

"You know, judging by their relative DNAs, the San Bushmen and the Hadzabe are about the same age," Wells says. "But I sent you to see Julius and the Hadzabe not only because, like most of us, you share their DNA, but because their lives are still more culturally traditional, in the hunter-gatherer sense. The San will put on a hunter-gatherer show for visitors, but they've been largely modernized; when outsiders live with them in their traditional manner, it's more of a commercial transaction. It's like theater. With the Hadzabe, if you go there and live the hunter-gatherer lifestyle, it's 100 percent real. They're among the last true hunter-gatherers in Africa. They're still living much like they did 25,000 years ago."

So, on the morning after our visit to the baobab tree, Julius arrives in my camp again, having already sent the rest of the tribe's men off to hunt. "This morning I will show how we gather," he says.

Minutes later, we join the clan's women and children, traveling to another bone-dry riverbed with fruit trees lining its banks. Over the next hour, we fill the clan's sole aluminum pot with small, piquant, juicy red berries. As we work, Julius snaps off small twigs from a riverbank shrub the Hadzabe called *tafabe,* but that scientists call *Salvadora pesca*. This is Africa's "toothbrush tree," and everyone scrubs their teeth with the fleshy end of a broken twig as if using a toothbrush. Given their diet and lack of modern hygiene, you'd think the Hadzabe might have terrible dental health. Instead, their teeth are strong, white, and healthy.

When I ask about this, Julius pauses from scrubbing his teeth and chuckles. "Our secret?" he says. "No sugar."

After about 40 minutes, with the pot filled with berries and everyone's teeth sparkling clean, the women next brandish pointed wooden sticks about three feet long and begin walking slowly up the dry river's banks. In about a minute, one of the women points another 20 feet down the river, to a spindly stalk of plant about four feet tall with large green leaves. "This is what we call *shuma,*" Julius says. "Now watch this."

The women drop to their knees and slowly, gently begin to dig into the earth around the plant.

"They're looking for tubers," Julius says. "The women won't disturb the *shuma*'s roots; this is important, so it can shelter more tubers in the future." In 40 minutes, after exposing nothing but dust and cobble, the excavation finally arrives at the *shuma*'s business district, behind which comes a chain of linked brown tubers, each the size of a football. With the appearance of each new tuber, the women grow happier. "This evening," they say, "we will feast."

Having spent several days with the Hadzabe, my overwhelming impression is that they are among the most relaxed, self-assured people I've ever encountered. They have complete mastery of their landscape. For them, on most any day, only today exists: warm or cool, belly full or searching out a meal to enjoy. When the work is done, they explore new ways to shoot arrows, or find a better piece of wood for starting a fire, or boil different leaves, berries, and roots to make teas or soups or medicines or dyes. Certainly, there is the nagging fear about the future, about the incursions of other, more desperate tribes, but those don't seem to occupy too much of their daily lives.

Among the Hadzabe, I feel a peaceable happiness I've only experienced once before, when I spent several weeks among the Yanomami of the Amazon: a native people that, not incidentally, are also hunter-gatherers who live in relatively small groups and are seminomadic.

When I tell this to Julius, he chuckles. "We have been here more than 10,000 years, maybe ten times that long," he says, clicking away. "So yes, we understand this place. We are comfortable here. This is the meaning of home."

Too soon, it is time for me to begin to push on, leaving the Hadzabe for the next arm of this tour of my extended family: the Middle East. Still, before I depart, Julius gives me one last insight into what makes his people—and all human beings—unique and remarkable in the world.

A few days earlier, we'd walked past a small local school, the one built three years earlier expressly for the children of the indigenous Hadzabe. Now, it is mostly populated by children of the few Masai and Datoga herders in the area, and I noticed lots of Masai children dressed in the characteristic red-and-black checked clothing of the tribe inside, as well as many children from the herding Datoga tribes that have moved into the area, but those children are dressed in more Western-style clothes. Looking around, I could see there were only a few Hadzabe kids.

On one of my last days among the Hadzabe, as we walk back toward Julius's camp, we pause again to stare at the school for a minute, watching from a distance. The building stands beneath tall shading, deciduous trees, watered by a mechanically pumped well. Nearby is a second building, a long, low-roofed warehouse. The ground is powdered with dust, but the kids look happy and thriving. Some are eating fruit off a tree in the schoolyard. In the late day sun, the scene is beautiful.

Given that this is Hadzabe land, I had to wonder: What was up? I ask my host, "Where are the Hadzabe kids?"

Julius sniffs. "This is a boarding school," he says. "Most of the children here sleep overnight in a dormitory. We like to have our children with us at night. We like to feed our own children. Still, our children do attend this school. But they have to stay outside; they don't go inside the school itself. The school roof is made of corrugated steel. The dormitories have steel supports. Remember: We are natural, we don't live our lives inside of steel structures. We cannot sleep beneath a metal roof. That is not our way. This creates difficulty, since we know that education will almost certainly give our children many more options in the future, and that is something we want to give to them."

Julius pauses. He stares at the ground. He kicks the dirt.

"As chief, I have a growing problem," he says. "It's a big problem. The Masai leave our conservation lands at night, taking their livestock with them. They go back to their own settlements. But the Datoga don't always do that, and we are not happy about it.

They bring their relatives and livestock from their own lands in the Ngorongoro Highlands, and they now have washed across our valley. They eat our fruit and drive off the animals we hunt. We have complained to the government about this many, many times, but the Datoga are better educated than we are. They are more popular with the government than we are. They are more populous than we are. They have more resources. Now they will stay, I think."

Julius pauses again, the focus in his dark eyes jumping around the landscape, nervous as a bird. Finally he says: "But that is their side of things. Meanwhile we keep living with our old tribal ways. We keep on with the life we have always lived: hunting and gathering and living peacefully in nature. This life leaves us more isolated every month as the Datoga influence grows on our land. At most, there are 1,200 to 1,700 Hadzabe left. I fear for us in the future. I fear for us. As the chief, this is something I think about all of the days."

He stares at the ground between his feet, and then at the school, then back to the ground. His gaze now seems to be boring a hole into the powdery, dusty earth. "We live simply, but we are not stupid," he says. "Life is change, I guess."

But on my last night in camp on Hadzabe lands in the Rift Valley, Julius also shows that he's developing his own flair for adapting. As evening falls, I am sitting in a wooden folding chair in my own camp enjoying a large, African-style bottle of Kilimanjaro Lager. It is among the last in my cooler, a box that, after a long time in the bush, is now bereft of ice. And in the near dark, Julius walks over from the Hadzabe's camp, 300 yards away.

When I glance up and see him coming, he lifts a hand. As he approaches, he is smiling. Robert comes over, to translate. As Robert walks up, Julius points at the beer bottle in my hand. "Do you have another of those?" he asks.

"Sure."

"I'd like one."

"I have other drinks, too," I say. "Coca-Cola. Bottled water. Orange Fanta. I can make you some tea, if you'd like. There's no more ice in the cooler, so the bottled drinks are not so cold as they might be."

"It's okay."

"So what would you like?"

Julius smiles and points again at the big, brown-glass bottle in my hand. "I'd like a beer," he says.

I open the cooler and rummage inside, looking to fill his order. It is then that one of Julius's lessons of the last few days hits me.

"Julius," I say, "you understand that this beer was once stored inside metal vats. It was brewed inside huge metal tanks. It was pushed around with mechanical pumps. This beer required petroleum products to be manufactured. It is not natural." I smile.

Julius pauses for a moment, then shoots a quick smile back. "Oh, yes, yes, I know all of this," he says. "But I have been to the towns. I know somewhat of the bigger world. There are some things about it that I like. And even a beer that is not cold is a rare treat."

With the metal opener of my Swiss Army knife, I pop the cap off the beer and hand the bottle to Julius, setting the knife down on the wooden folding table that stands between our two chairs. He accepts the beer, nods a quick thank-you, and takes several long, luxurious slugs from the bottle.

Finally, he starts talking. "Even as a Hadzabe chief, it is important to recognize that a man has to be a little flexible sometimes," Julius says. He smiles again. "I guess this is one thing that has always been true for the advancement of human beings. . . . For me, I like to think breaking the rules sometimes helps to keep me human."

We sit another few moments in silence, Julius looks at my Swiss Army knife on the table between us. With a lifting of his left eyebrow, and a finger pointed toward the knife in a bit of pantomime, he asks if he might lift it up.

I nod, and flip my hand, palm up. "Yes, sure," I say.

In the now near darkness, Julius lifts the knife from the table between us and begins to play with its dozen or so different tools: experimenting to see how it might make his life a little easier.

A few minutes pass. We sip our beers quietly. As we do, Julius remains focused on the knife, interested in exploring all of its possibilities.

After that, we sit for a long time, watching the last of the day give way to the sparkle of the evening's first stars. Birds flutter in the trees. Somewhere in the distance, Julius's dogs bark. When Julius and I don't talk for a few more minutes, Robert excuses himself, asking that we call him should we need any translation. We don't really need translation anymore.

*A young Arab girl, another relative from a long-ago branch of the ancestral family tree, outside a shop in the bazaar district of Baalbek, Lebanon*

# LEBANON

**WHEN MY FLIGHT LANDS** at Beirut International Airport two days later—one of my duffel bags, lightened by one Swiss Army knife and one Petzl headlamp but enriched by several pieces of beaded jewelry made by the women of Julius's band—it's as if, in the course of a day, my existence has leaped 25,000 years forward.

Everything is different. In the sunny morning glare of the airport's International Arrivals Terminal, a white automated teller machine dispensing both U.S. dollars and Lebanese lira stands like a modernist sculpture near passport control, ready to help cash-tight visitors extract money and purchase entry visas. Arcing white beams of steel support enormous and spotless windows, through which the daylight streams. Despite the humid heat of the Middle East in late June, the terminal's environment is air-conditioned perfection. On the far side of the enormous immigration and baggage-claim hall, heavy doors slide open regularly, and beyond them a busy coffee bar and newsstand await. As people stand in line for visas and passport approval, several of them check their Blackberries and Trios for e-mail updates.

I am not so busy. In fact, I'm consciously trying to avoid being busy. For almost two weeks, I've been living among hunter-gatherers, and I want to preserve the easy, swinging pace of that life as long as I can. Despite what Dr. Wiessner told me about the Hadzabe "maintaining social order" all day long through conversation, my time with them was among the most relaxing and pleasant windows of days I've ever been lucky enough to enjoy.

Suddenly deposited back in the modern world wearing my dirty clothes and still feeling the placid glow of isolation, I'm actively blocking an impulse to check the voice mail on my mobile phone. It's fun to relish the changes of 25,000 years in a couple of airplane flights. Now, beneath my feet, instead of powdery and ash-gray dust, my moccasins are planted on a polished floor. The floor alternates in meter-a-side squares of black and white marble, creating a chessboard motif. In Lebanon, as we will learn, the metaphor is appropriate.

I've been told that, just beyond the doors of the secure luggage-collection area, people will be waiting to assist me. A gleaming black BMW limousine is to be there, too, to hustle me along to the fashionable Le Meridien Commodore hotel downtown.

Is this really possible? Can a hunter-gatherer existence and a modern one really be spliced together by a couple of airplane rides? Is it really possible to leap 25,000 years into the future in just a little over one day?

Thirty minutes later, the promised limo delivers me to my hotel. Located in downtown Beirut, with a monolithic white exterior punctuated by large, equal-sided and heavily framed picture windows reminiscent of the old "Hollywood Squares" TV set, the Commodore is up-to-the-nanosecond modern . . . and absolutely accommodating. Just inside the front doors, at the reception area, I'm offered the choice of a cool welcome drink or a cup of tea; then told to relax as the desk manager checks me in.

When I present my Le Meridien member card, the manager offers me a free room upgrade, like I'm an old friend. As the paperwork is being completed, I relax in the otherwise unpeopled lobby. Furnished in various types of varnished wood, with sleek modernist tables and chairs, and lit with recessed spotlights and bulbous chrome-plated lamps that hang from ceiling beams and work to create dramatic pools of light on the floor (despite the morning sun), the lobby is yet another air-conditioned haven in modern Lebanon: a welcoming place to relax. Just audible is a techno-mix of ambient music not edgy enough to be rock; almost jazz, and it unobtrusively fills the air. This place, with its self-consciously minimalist modernism, is a perfect thumbprint for fashionable hotels of the postmillennial age. In fact, these hotel lobbies with their low-slung furniture and discreet bars and ambient soundtrack wrap today's world from New York to

Bangkok as surely as sunshine. This place is also light-years from the dry earth I slept on in Tanzania only a day ago.

The tea arrives. I extract my laptop from my briefcase and fire it up for the first time in almost two weeks. The lobby, of course, has wireless Internet. With the click of a few keys, I log into my primary e-mail account to find a two-week backlog of messages that number in the hundreds.

These—it is immediately decided—can wait.

In a matter of hours, I've been slingshot from prehistory back into the modern world; into a human existence equal parts ease, comfort, anxiety, and distraction.

A small folder with my key cards inside arrives from reception, carried over by the desk manager, who is nothing if not understated. With a single swipe of a credit card, I've been granted a temporary, and very pleasant, new home. There is a pool and a gym, the man tells me as he hands the keys over. There are three restaurants. He hopes I'll enjoy my stay. A porter, he informs me, has already swept my duffel bag from the front desk to my room.

The folder with the key cards goes into my pocket. I slip my laptop back into the briefcase, sit back in the comfy chair, and sip my tea. Much as I enjoyed my time with Julius and the Hadzabe, the modern world has pleasures to recommend it, too. And about now I'm ready to enjoy a few.

My room, on the hotel's seventh floor, is clean, bright, and decorated sumptuously: with white walls, Swedish modern desks and night tables, and crisp white linens overlaid with a textured brown spread that falls precisely across the bed's horizontal middle third. I flip on the TV. CNN is there: People are shelling one another somewhere. Others are running at night through a city in flames.

As the TV plays, I walk to the room's far, exterior wall, where the sealed, floor-to-ceiling, *Hollywood Squares* windows bring the city and its bright daylight inside. Outside stands a robust garden of modern glass buildings and skyscrapers, the whole scene set off against the gorgeous blue of the Mediterranean Sea.

I stare longer, listening to the news on TV. In another 30 seconds, closer inspection of the city begins to reveal a landscape that, at first glimpse, was not apparent. Beirut is a jumble of small roads that intersect, diverge, and then, coursing the hilly topography, turn back, sometimes to intersect again. Along these small and probably ancient thoroughfares, in rubble around the new skyline's knees, are older buildings, often of whitewashed plaster, many with red-tile roofs. A surprising number of these older buildings are shattered, their tops fallen in. Some of these smaller, older buildings have huge, gaping holes in their sides; exterior walls that appear to have been punched in years ago by explosives, and have now been left to develop a patina of dusty neglect that implies not only entropy but also a lack of rebuilding resources and the slow but relentless tug of gravity.

It's understandable, I guess. Since modern Lebanon emerged in the wake of World War I, where it came to survive under

French control, having been wrested from the crumbling Ottoman Empire, it has ridden a series of golden ages, often abetted by its Western-leaning politics and economy.

These periods, however, have been regularly punctuated by internal and international strife . . . often characterized by religious tensions coupled with regular rocket attacks and explosions. After France released Lebanon to self-rule at the end of World War II, the initial government rested with the nation's conservative Maronite Christians, with the nation's majority Islamic population largely excluded from both political and economic power.

As day follows night, especially in the often fractious Middle East, this arrangement soon led to chronic violence, which has sometimes been fanned into flame by Lebanon's southern neighbors, Israel and Palestine, and its eastern one, Syria. This has only become truer as, since the 1950s, Palestinian refugees began moving into Lebanon's more peaceful southern districts, which sit only 35 miles or so away from the former Palestinian territories. It's an in-migration that's imported Israeli-Palestinian violence cycles into Lebanon, until regular violence became a fact of Lebanese life.

Finally, in 1978, the burden grew too great: The Israelis, claiming a need for security from Palestinian cross-border incursions and violence from Lebanon, marched less than peacefully into southern Lebanon and set up an Israeli surrogate militia force, the South Lebanon Army, with the rationalization that they had to protect themselves with security on both sides of the border.

While international pressure drove Israel from Lebanon a few months later, invasions into Lebanon by its neighbors have come regularly ever since, destabilizing the delicate balance of Lebanese self-rule. Even the Syrians have gotten into the act, regularly crossing into Lebanon to "help establish peace."

All of these unwelcome visitors have inspired both the Islamic Lebanese and the Christian power bases inside the nation to arm themselves more and more heavily: Weapons they sometimes feel compelled to use on one another in a low-simmering civil war, with everyone trying their best not to acknowledge the tensions while also trying to live, raise families, and enjoy this place and its history.

By the 1980s, Lebanon's large Shiite Islamic sect, wanting to better look after its own and protect itself, formed Hezbollah, which, depending on whom you ask, is either a political party and social-development organization . . . or a terrorist group.

Middle Eastern politics rarely grow less complicated over time.

To experience some of this for myself, later today my plan is to walk the streets of downtown Beirut, headed for the wide sidewalk of the seaside Corniche Promenade, against the foaming Mediterranean surf that smashes against the Corniche cliffs. Eventually that route will pass a block of rusted and still wrecked hulks of exploded cars and trucks—a blast zone a few years old—in front of the once stately St. George Hotel. There, on a barricade-closed street called Rue Minet al Hosn, back on February 14, 2005, the former Lebanese Prime Minister Rafik Hariri was in a motorcade on its way to an official event when an estimated 2,000 pounds of TNT was strategically detonated, killing Hariri and 21 members of his motorcade.

His assassins, whether Lebanese, Syrian, Israeli, or Palestinian, have never been identified and brought to justice, though the investigation has gone on for years and is said to implicate Syria. Still, the Hariri assassination solidified one aspect of Lebanese politics for the foreseeable future: Intrigue in these parts is so thick that all sides can logically blame the explosion on someone else, with no consequences ever arriving. And yet the fact remains: Someone actually did it. Car bombs with the equivalent of a ton of dynamite don't just materialize and detonate with such devastating timing.

In spite of this terrible event and the subsequent investigations, nothing has been done to quell the violence in Lebanon. In this environment of mutual distrust and anonymous aggression, it's no wonder suicide bombings, international incursions, helicopter gunship attacks, and internationally fired artillery shells regularly threaten the country. For years now, because of the random and constant threat of violence, the U.S. Department of State has strongly and publicly advised Americans against traveling to Lebanon. The violence is too random, the State Department says; the source of danger too hard to quantify.

This too is a tragedy, because the Lebanese nation, with its snowcapped stony mountains draped at the lower elevations with tiered olive groves, the stunning blue of the Mediterranean, and a climate that's usually—if sometimes steamily—pleasant, is delightful. And its people in their individual dealings with visitors are delightful, as well.

The food's pretty spectacular, too.

Still, as I stand in my hotel window, overlooking the mirrored skyscrapers of Beirut, with the neglected, explosion-shattered buildings of the city at their feet, the unresolved hard feelings and destruction are no easier to comprehend. But knowing the history, the scene is at least a bit easier to grasp.

One thing remains certain, however. I'm a long way from my friends the Hadzabe. Back in Tanzania, despite the Hadzabe's own catalog of real and pressing worries, I could at least feel confident nothing was going to explode unexpectedly. This is a truth that, in the end, at least allows a person to sleep a little better at night.

Which was one of the strangest yet truest lessons gleaned during this whole trip. The closer you get to cutting-edge, contemporary life, the worse you sleep.

My ultimate destination in Lebanon is the Bekaa Valley, a broad, rich farming plateau located between two 10,000-foot mountain ranges, about a 90-minute drive from Beirut. There, Spencer Wells informed me, an indigenous Lebanese Arab population that's been there for millennia is the next group with which I share a genetic marker. This marker, called *M89,* turned up in the human genetic code about 45,000 years ago, and has been living here ever since. Once again, just like the *M168* marker that I, and most males, share with Julius and the men of his band, *M89* is also very common, tracing one of the classic migration routes out of Africa. Also, unlike my inability to

*The last standing pillars of the Temple of Jupiter in Baalbek. At more than 72 feet, they are the tallest Roman columns on Earth.*

speak click-talking Hadza, I can speak a little Arabic. I don't want to miss a thing, though; again, I'll need an interpreter.

So on my first night in Beirut, with the help of a local friend who is a producer for CNN, I'm introduced to Maya. A receptionist at a Beirut beauty shop, Maya is fluent in Arabic, English, German, and French. And she's looking for a working vacation; a few days away from her usual desk.

When we meet in the hotel lounge, she tells me all about herself: She was married for several years to a wealthy German businessman and traveled and lived with him in several of the capitals of Europe. Still, in the end, she came home. "It was just better," she says. "I'm just glad to be home now. I missed my family. Missed my friends. In the end, my ex-husband and I just weren't completely compatible."

With sunset glowing outside, we settle on a per-day price for her translation services over the upcoming four days, and agree that I'll pay for her meals and hotel. Then, business finished, we step to the concierge desk and order a car to drive us to Baalbek a little after 10 a.m. the following day. Maya says she'll be back tomorrow, bags packed and ready to go.

The next day, at about 10:30, after a cup of coffee in the lobby, our driver arrives, and we're off. The drive is a dramatic one, and shorter than I might have thought.

That's another thing about the Middle East. Everything is close together. This is especially apparent after recent travels in the United States . . . or the plains of Africa, where things are often great distances apart. In the Middle East, though nations may be millennia apart in terms of understanding, they often sit, boundary to boundary, as neighbors.

Despite being the national capital and a city of two million, Beirut proves easy to navigate, especially once we're away from the crowded, narrow streets of the downtown Hamra District, where the hotel is located. Still, as the driver navigates through the city, exceptional wealth and threadbare poverty press tightly against one another, accentuating this country of contrasts. We pass a crumbling building, its roof tiles fallen inside the building's whitewashed husk; its windows broken out, their frames staring at the looming high-rises and their rooftop satellite dishes. As soon as we get away from the

Hamra, things settle into the old Middle East quickly, and up a side street I glimpse both goats running on a hillside and a two-wheeled cart being drawn by a donkey. Within a few miles, the roads grow wider, leading to a central artery routed away from the city and into the foothills of the steep Lebanon Mountains, now only a half dozen miles to our east.

The car is soon climbing steadily upward on this good road, one of the country's main superslabs, quickly making its way up the side of the 10,000-foot Mount Lebanon, leaving behind the glint and glare of Lebanon's Beirut/Mediterranean side for the Bekaa Valley, just on the other side of the mountains.

As the elevation rises, the temperature changes quickly. Olive trees are replaced by the gnarled, famous cedars of Lebanon: A tree so hardy and long-lived it figures as the central image on the national flag. Before long, in shaded places, bits of snow cling to the mountainside. We open the car's windows, and instead of encountering the steamy Mediterranean heat of the last days of June, we're greeted by a cool and dry wind pouring into the car.

It's beautiful. Refreshing. And instead of feeling a sense of slightly anxious anticipation about our destination, as I did back in the hotel (the Bekaa Valley and Baalbek are the birthplace of Hezbollah, after all), there's a growing sense of renewal and relaxation. As we crest Mount Lebanon through a little notch in the ridgeline where snow covers the ground on both sides of the road, the world spreads below us like a green and gold quilt of seemingly endless cultivated fields.

At first glance, the Bekaa Valley is gorgeous. Hemmed in by the mountain range we have just crossed and, 15 miles away, the equally tall Anti-Lebanon Mountains, which define Lebanon's border with Syria, the Bekaa Valley is perhaps 75 miles long and is well watered on both east and west sides by icy snowmelt rivers and cool springs that run down off the mountains year-round.

As quickly as we ascended the mountains, we're down on the valley floor and soon rolling into a little village of trim houses and small streets. Along a central river, the town seems virtually overrun with outdoor eateries: restaurants and food stalls. Unlike the dusty tumble of Beirut, which we left 45 minutes ago, this town is calm and almost placid. Instead of shattered buildings, there are flower gardens and vegetable plots. It's like a completely different country.

"So . . . this is Zahale," Maya says. "It's a favorite place in summer; many people come over from Beirut just for meals, because the weather is so pleasant and you can eat outdoors. Which you can't do with the heat of Beirut in the summer. You'll see. The Lebanese people love to eat outdoors. It's an old, old local custom. Really. It sounds funny, I know, but in summer people really do come over here just to eat. It's pleasant . . . and beautiful."

We follow the road out of town to the northwest, the car hurtling past old trucks and motorbikes at spots where we can pass. Even as midday approaches, the air stays much cooler and less humid than in Beirut, though we're only about 20 miles from the Mediterranean and its sea-level smolder. We're in a new climate: This valley hangs between the two mountain ranges, several thousand feet above sea level.

The car keeps speeding smoothly on, and I lose myself in the scenery: vast farmlands crossed with ample irrigation canals manned by farmers directing the water to different fields and areas where livestock has been penned. Some people ride donkeys and others lead pack donkeys along the road on rope halters, the animals' backs heaped with cloth-covered bundles.

Farther to the north, outlined against the brown eastern slope of the valley floor, there's an unusually thick patch of tall and ancient-looking trees: big pines and cedars. And there, scattered among the tiered levels of the far mountainside, interspersed between the trees, stone columns and soaring architraves slowly come into view. These are the remnants of ancient Roman temples, and amid the greens of the trees, they stand in counterpoint to the white squares of small mountain houses and the muted browns of the mountainside.

Up ahead is Baalbek.

A well-trampled city with a truly prearchaic history, Baalbek was named for the Phoenician god Baal (who, among other things, was considered a maker of weather and who stands as an Old Testament competitor to the notion of Jaweh). Later, following a period of Assyrian control, Alexander the Great conquered the city in 334 B.C. and rechristened it Heliopolis: "City of the Sun."

Whatever the motivation for renaming the city, not to mention the eventual slide back to Baalbek, during the nearly three centuries of Hellenistic control, the Greeks moved the local culture onward along with the rest of their society across the Mediterranean. This included the celebration of religious rites on a natural ridge overlooking the valley below, with the art-directed, stage-symbolism of such a location being lost on no one.

On this low, flat ridgeline halfway up the side of the western range of mountains and overlooking the valley, these religious celebrations even came to be performed on an official podium. Eventually, the podium's stone foundation would provide a locus for the now crumbled temples constructed by the Romans, who followed the Greeks into this valley in 64 B.C. Unwaveringly knowing a good thing when they saw it, the Roman conquerors not only took the city from the Greeks for esthetic and agricultural reasons, they kept the name and made the entire, sun-and-water-kissed valley one of Rome's granaries for the next four centuries.

Across the 400 years of Roman rule, the Bekaa Valley prospered as never before. Enormous temples and courts were constructed in stone, much of it taken from nearby quarries or imported from areas of historical significance or geological uniqueness. Within a few generations, Heliopolis would become the pillar of the Roman Empire's farming hinterlands; Rome even christened Heliopolis the center of worship for the sky god, Jupiter. From this valley, fields of grains and groves of olive trees poured their bounty onto ships bound for peoples and legions of a conquered world that spread from England to Spain to North Africa and the Middle East. For all of these places and peoples, the bounty of Heliopolis played an enormous part in the force that was ancient Rome.

But in centuries to come, Baalbek and the Bekaa Valley would also welcome more conquering peoples. With the fall of Rome came the Byzantines, followed by the Arabs, the Crusaders, the Ottoman Turks, and the French. And with each new wave of conquering armies came yet another layer of foreign influence and character built upon the valley's already rich past.

Today, with roughly 300 sunny days a year, some still fertile soil, and ample water, Baalbek and its Bekaa Valley are among the most beautiful and history-infused places a visitor can seek out.

Occupying the highest point in the Bekaa Valley, and protected between the steeply majestic walls of the Lebanon and Anti-Lebanon Mountains to the west and east, plus the high plateau bench just to the city's north, its obvious why every culture that's arrived in Baalbek and the Bekaa Valley has found it to be a virtually impregnable fortress with good vantage points in every direction, not to mention a perfect locus of agriculture and activity.

And as we roll into town, activity is everywhere. People are working in the fields. And donkeys and a few camels, their flanks laden with woven baskets of vegetables and greens and reeds, are being led along streets and pathways. The roads are choked with trucks, their cargo beds freighted with potatoes and sugar beets, onions and other row-crop produce.

Though it's not yet noon, the day is already hot, and dark, heavy, saturated shade hangs straight down beneath spreading trees as we enter town along Rue Abdel Halim Hajar, the central market street. Near the edge of Baalbek, on the right-hand, eastern side of the road, backed against a tree-shaded hillside and tucked behind an ivy-covered stone wall, is a hulking old building: a massive edifice of sun-washed stones. The car slows in front of it, stopping precisely at its iron front gate.

"Here we are," Maya says. "This is the Palmyra Hotel. You'll like it, I think."

We step from the car into the day's dry and withering heat. As we do, the driver is already out and bustling around, pulling our baggage from the trunk and placing it on the sidewalk next to the hotel's old stone walls. Clearly, he's ready to get back to Beirut. I pass him a tip for his quick work . . . and in a cloud of dust, executing a quick U-turn, he's back on the road and gone.

Seeing the activity in front of his building, the hotel's manager comes out the opened double front doors and into the sun. Against the hotel's two-story stone face, all of the green-painted wooden shutters and unscreened windows are opened wide; a welcoming visage.

"Hello . . . hell-o," the manager says. He's smiling. The manager is sort of round and bald. He's dressed in pressed gray trousers and a white button-down shirt with the sleeves rolled up. He's wearing an unbuttoned vest whose fabric matches his trousers. He lifts his arms, opening out his hands.

"Welcome," he says. "Welcome to the Palmyra Hotel. I know you will be happy here. I have already called the porter. He will soon bring your bags to your rooms."

We follow the manager past graveled courtyards on several different levels and an array of flowering shrubs, ultimately passing through ancient-seeming, glass-paneled doors. And once inside, the place seems to yawn open. The central salon is huge, a two-story-high cube paneled in dark wood, with enormous and ancient-looking tapestries hanging from the walls and, to the left, a line of worn stairs leading up to the second floor. On built-in shelves surrounding the salon are rows of books and ledgers and the sculptural busts of men I don't recognize. Long wooden benches and chairs line the walls of the salon. The place is dark and a little dingy, yet very comfortable. Sun streams in through the open doors and windows, making bright panes of light on the Oriental rugs and wide wooden planks of the floors. Even in the now stiflingly still air, thanks to its patina of age and "gothic house of horrors atmosphere" (cited in one of the guidebooks), the Palmyra, built in 1875—with its old wavy-glass double-hung windows and shutters to be closed at night—coolly evinces character. I love it.

Standing in the main salon and watching through the front doors as a porter begins loading our bags onto a dolly, the manager asks us to follow him into his office, a small room to the right of the grand salon. Within, antiques and framed images cover the walls. A computer sits incongruously on an antique French desk.

"Please, please, sit if you would like," the manager says. "Would you like a cup of coffee? Tea? If you will give me your passport and a means of payment, preferably a Visa card if you have one, I will take care of the rest. I will also need a copy of your passport, to give to the police here. But this is only a formality. It's nothing really, just something that must be done with all foreign guests. That's all."

The manager takes my credit card and passport, switching on a dormant copier near his desk. As he waits for the copier to warm up, he tells us a little more about the hotel. As the first Western-style lodging ever constructed in Lebanon, it has witnessed world events over more than a century. During World War I, it was a barracks and command base for the German Army. In World War II, the British used it as their regional headquarters. Charles de Gaulle slept here—in Room 30, it turns out.

"Would you like Room 30?' the manager asks me.

"It doesn't matter," I say. "Any comfortable room is fine."

According to the manager, Jean Cocteau was a huge fan of the hotel, too, and he left behind drawings done here that can now be found displayed on the walls.

As the manager tells me this, he shuffles through the room keys, each of which has a heavy metal fob attached. The copy machine beeps: ready for action.

"I have planned for you two very good rooms, though in different wings of the hotel," the manager says. He hands us each a key, and gives us rough directions to our chambers. Mine is up the stairs to the right; Maya's up the stairs, down a hall, and to the left. He makes a copy of my passport, then hands it back.

The manager smiles. "Okay, there," he says. "It's funny, this week is very quiet. Next week and across the middle of July, we have a cultural festival here every year, with theatrical plays and symphonies playing in the old Roman amphitheaters and ruins. It's quite beautiful. But that doesn't start until after you're scheduled to leave. And, anyway, the hotel is full for that time. Though, if you wanted to stay, I'm certain we could find a place for you somewhere here in the city. In the meantime, however, I hope you will be happy. And please, if you need anything, just ask."

Within a few minutes, our bags deposited in our rooms, it's time to explore. The rooms are simple but comfortable, with ample light, old armoires and chests of drawers; there are no TVs, but there are enormous bathtubs, which were installed by the British in 1924.

By the time we're settled in, it's about 1 p.m., and I'm ready to explore. Maya says she'll join me on a quick tour of town, but then she wants to return to the hotel and relax a bit in her room. So after checking out the second-floor landing and its ganglia of hallways, each of which is hung with still more tapestries and paintings, Maya and I head out onto the street, where the day is now suspended in sun-blasted, midday, Middle Eastern torpor.

Once again, I'm ready to meet more of the family.

Once we have passed the tree-shaded gardens and terraces and the small reflecting pools inside the hotel walls, just beyond the Palmyra's front gate, the city seems to sleep. An hour ago, cars were coursing up and down Rue Abdel Halim Hajar. Now, it's empty and seems to be dozing through the hypnotizing quiet of a summer afternoon. Just past the city's first fountain square there's a shawarma stand; its spitted meat turning and browning for the hungry to see. The door and windows of the shop are open to the street, but no one is available behind the counter to carve meat for a sandwich; they're probably in back, dozing. A few doors farther along, on the other side of the street, a barber naps in his chair, the lights of his shop turned out and the shop's door wide open. The midday heat seems to have crushed everything, stopping the town center in mid-activity.

Up the street, about where the pavement turns to deeply set cobblestones, several men in linen or knit skullcaps and linen robes sit on the shady side of the street on small wooden benches, soporifically drinking tea, smoking hookahs, and conversing slowly and quietly. As Maya and I approach, they look away.

"*Salaam alekum*," I say, wishing them the traditional Islamic greeting of "Peace be with you." They nod hello, saying nothing in return. As we walk past, I lift my open right hand. They're neither friendly . . . nor unfriendly. They're just doing their thing. Along the street, I've noticed that many shops have caged red or gray finches hanging outside their doors. In the dry heat, the birds chirp quietly in little high, chippy sounds; they hop nervously from their perches to the cage's bars and food containers.

We keep going. Finally, I lock eyes with a man seated on a burlap sack containing coffee beans in the shade of his shop along the sunny side of Rue Abdel Halim Hajar, which has now become a very narrow street.

"So, . . . " he asks, "do you belong to CIA or UN?" He holds up his hands in mock surrender and laughs. "No one else comes out in the heat of the day. And you are new to the town. I have never seen you before."

I chuckle too, stepping into the shade of his shop. "I'm neither," I say. "I'm just here for a visit."

Maya steps in out of the sun, as well, and begins looking around the shop. I look around, too, and my eyes find a freezer near the cash register. Inside is everything from a Popsicle to a Drumstick to various kinds of ice cream in containers. I extract an ice-cream sandwich from the freezer, and ask: "How much?"

"Seven-fifty," the shopkeeper says. That's about 80 cents. We make the transaction.

Maya and the shopkeeper have a quick conversation in Arabic that moves too fast for me to understand. They smile and make hand gestures. I unwrap the ice-cream sandwich.

"Cool off, stay cool," the shopkeeper says as I peel back the paper. "This is Baalbek. This is a place where we always try to stay cool."

Not too much farther up the sidewalk, two Arab men, dressed in gowns, sandals, and full red-and-white houndstooth Arab kaffiyeh headwraps, motion for me to join them as Maya and I leave the shop, my ice-cream sandwich already half-gone, "Hello, peace be with you," they say in Arabic. "Where are you from?"

"America."

"Good, good America!" They nod and smile. "How long do you stay in Baalbek?"

"Three days."

The older-looking of the two men smiles. He has a heavily lined face that's stippled with a few days' white whiskers. His hands are wide and thick. He and the other man are playing chess in the shade. "There was a time when Superpowers came to Baalbek and stayed for hundreds of years," he says. "Phoenicia, Greece, Rome . . . now America comes for three days."

Afterward, he sort of smiles. It's impossible to tell if he is criticizing current U.S. policy in the Middle East or making a historical observation. Only he knows for certain.

I ask him his name.

"No names," the older man says.

"Then how old are you? What do you do here?"

"I am 76. I own this fabric store. I have been here with this store for 45 years. Before that, I was a farmer out in the valley. I was in agriculture, planting the land: growing wheat . . . corn. But I didn't own the fields, and the owners eventually took them back from me, and so my future was gone in the fields. Anyway, I prefer this shop to the fields. This shop has given me a life. I have had two wives. The first one died, which still makes me sad when I think about it. This shop and its life has provided for my ten children. Thirty-five grandchildren. Do you have a wife? Children? Grandchildren?"

"I have a wife and two children," I say. "But I have no grandchildren yet. My children are still young, teenagers."

"More!" he says. He shoots a fist in the air and smiles broadly. "You need more children!" he shouts. He waves his fist high in the air. As he does, his left knee slightly upsets the chessboard between him and the other man. "Make the world!" he half-shouts. He smiles.

Across the street, another man, perhaps noting the recent volubility of our conversation, steps from the shadows of his general merchandise shop into the afternoon sun. He waves me toward him.

Squat and scowling, he looks like an interesting prospect. I wonder if he might be an *M89* family member of mine. He has dark eyes and a nose so prominent it dominates his face. As I cross the street, I extend a hello and introduce myself. He does the same. His name, he tells me, is Halim Taha.

"I don't know if we are anciently related," he says as Maya and I finish explaining my errand in town. "But I welcome you."

As it turns out, Halim Taha has been politically representing this district of Baalbek for 36 years—he's a sort of district mayor—and he seems extremely proud of the town's easygoing style.

"There have always been people occupying us from the outside," he says. "Most recently, the Syrians have come, and now they have again left. But we in Baalbek have always been peaceful people. And anyone who visits can see why we're so happy. The beautiful nature. The cultural inheritances. The sun. The clean air. I cannot imagine a life anywhere else. No one born here can imagine it. We say: 'Baalbek is me, and I am Baalbek.' That's all."

Then he invites me inside his shop, which is festooned with everything imaginable, from lamps to bowls to china tea sets. "You don't need to buy anything," he says, as he steps back inside and out of the sun. "But it is hot. It is only smart for us to stay in the shade during the midday."

It doesn't take long to absorb Halim Taha's—and Baalbek's—chilled-out approach to life. Maya returns to the hotel for a phone call and a rest, and as the day tilts slowly toward afternoon, I fall right in with Baalbek's pace. By about 3 p.m., the shops are reopening, and life returns to the town, with shop doors opening and brimming with fresh vegetables. Before long, the smell of baking bread fills the air.

At the main market, the suq just past the main cross street of Rue al-Ain, I buy an *International Herald Tribune* and a Coke from a news vendor's stand, then sit in the shade of a Roman ruin and read. Lizards scramble along the ruin's stone walls. Even with the afternoon's reenergizing activity, the city remains quiet. Soon, I realize, the only real noise I can hear in the afternoon's glare and hot breath, beyond the occasional roar of an automobile engine or the honk of a car or motorcycle horn, is the rushing of water. It's insistent, always there. And soon, I am fixating on it.

In the early afternoon heat of late June, after I finish my Coke and newspaper, I go exploring for the water noise. It doesn't take long to see that it's coming into this area from everywhere: torrents of snowmelt race toward this area in grayish white profusion, moving through a series of culverts, aqueducts, and concrete-lined ditches. The ruin where I was sitting was part of an old city center, and several different streams course together into a kind of junction pool nearby, before rushing along a larger canal toward the valley floor below. All of this water runs beneath the two-acre grassy area at the ruin's center: a feat of architecture and excavation that implies the typical complexity of Roman engineering accomplished by builders working nearly two millennia ago without any truly mechanized digging implements.

The people who came to occupy this place 2,500 years ago obviously had the technology of crop irrigation well in hand.

---

Some time roughly 19,000 years ago, thousands of years after *Homo sapiens* had completed its dominance of the Neanderthals, among certain groups of modern humans something strange and new began to happen. After millennia spent hunting and gathering, the practice of gathering took hold as an easier, more reliable, and steadier source of food. Eventually, it begat the short but important period of human evolution called the Mesolithic, or Middle Stone Age. It's a period that's difficult to put a certain starting date on, as it arrived in different parts of the world not so much as a page on a calendar but as a set of evolving practices: a response to external survival pressures and opportunities.

By the time my ancestors arrived in the Middle East, *Homo sapiens* had been succeeding mightily for tens of thousands of years: Humans were hunting successfully and procreating at a steady and geometrically growing rate, with successive generations creating at first a few more, and then many more, individuals to feed, clothe, and house.

But this success and growth brought with it new problems. As hunters, humans had now become too good, too efficient. They were figuring out strategies in hunting for food, and by this time, according to some students of the Mesolithic period, they had even started to generate language, a system of sounds and gestures that could be widely understood and used to express their plans. According to some scholars, hunting and anticipating the strategy of the hunt had even gotten these early humans thinking about tenses: past, present, and future. It was an amazing leap from the desperate, in-the-moment, smash-and-grab of the previous 20,000 years to a form of far more organized taking. And, consequently, these hunters were killing animals to feed the hungry throngs in numbers that, even given roughly stable populations of prey, soon became unsustainable.

It's not surprising that, about 11,000 years ago, as humans began to arrive in disparate locations around the globe, the largest and slowest-moving species of the big game animals begin to disappear in unison. This includes giant bison with a horn spread of six feet or more, plus enormous beaver-like members of the genus *Castoroides* that could weigh as much as 250 pounds, mastodons, and two varieties of woolly mammoth. To these large animals, modern humans, with their advancing skills and hunting techniques, were an extinction threat on two feet. To the large, slow, and most vulnerable beasts of the age, humans had become, in the words of the Australian biologist Tim Flannery, "the future eaters," the title of his book on the subject.

At various locations around the world, the woolly rhinos and mammoths would soon be gone, as would several species of huge marsupials in Australia and New Zealand. Humans were suddenly hunting on an industrial scale. In the Czech Republic, the bones of at least 1,000 woolly mammoths have been found in one location. At a locality called Solutre in France, home of the famed "Solutre point," an elegant Paleolithic spear point of a type that seems to have been perfected there, the butchered remains of more than 100,000 ancient horses have been unearthed at another single site. Across the American West are "buffalo jumps" and box canyons littered with the prehistoric bones of bison and deer that were driven there and trapped for later butchering, or that were killed en masse to be quickly butchered and eaten.

But this stripe of killing wasn't limited to big game. In their rapacious search for protein, Mesolithic humans stood ready to devour any of the world's low-hanging fruit. According to a recent article published in the journal *Science,* anthropologists at the Smithsonian Institution in Washington, D.C., and at the University of Oregon studied ancient garbage sites, called middens, discovered on the Channel Islands off the coast of southern California. Excavations in these refuse heaps, dated to more than 13,000 years ago, show that ancient humans in the area first hunted sea otters to extinction, with more recent strata in the garbage pile showing that, with the otters gone, *Homo sapiens* in the Channel Islands began to concentrate on sea urchins as a food source: The urchin population had exploded with the lack of pressure from its most significant predator, the sea otter.

As a food, otters are more nutritional than sea urchins. So for the human hunters on the Channel Islands, the quality of their food came to be degraded by the aggressive hunting of that food itself. "Human influence is pretty pervasive," says Torben C. Rick of the Smithsonian National Museum of Natural History, one of the article's authors. "Hunter-gatherers with fairly simple technology were actively degrading some marine ecosystems thousands of years ago, and these effects cascaded down through the entire ecosystem."

*Homo sapiens* was ascendant. Humans were learning to eat most anything they could get their mitts on. And they were very, very hungry.

But reading the article in *Science,* my big question wasn't about the fact that modern humans were eating their way into protein bankruptcy. It was this: How did these early humans find themselves out on the Channel Islands, some 40 miles off the North American mainland? Even if ice ages had locked up seawater, dropping the levels of the oceans, crossing 40 miles through what are today some fairly deep oceanic channels seems almost impossible.

Still, Rick's point is an important one. After generations of stumbling along, barely surviving, *Homo sapiens* were suddenly thriving with bloodthirsty glee, and they were degrading the environment in the process. Modern humans, as hunters, had specialized. They were hunting large animals or easy prey to kill and eat as a way to prosper and promote the race. And the bloodlines of many species on Earth may have been cut short by these fast, smart, new predators on the block.

As the Channel Island middens show, at the same time the early humans were prospering, they were also painting themselves into a protein-hungry corner; trapping themselves through a one-sided hunting technology that would ultimately leave them with decreasing, rather than increasing, options.

Getting good at taking easy protein would soon prove to be humankind's first quantifiable technology trap.

---

Technology traps? What are these? As we'll see from examples presented later in this book, technology traps are paths of technical or practical progress in human civilizations that ultimately contribute to problems for these cultures—and in several cases, their destruction.

While the scientist, author, and thinker Jared Diamond has posited in his book *Collapse* that some cultures disappear as a result of environmental degradation, climatic changes, hostile neighbors, a lack of societal support, or an inability to adapt to new situations, I'd suggest that, in fact, all of these are symptoms of a larger problem.

That problem is cultural overspecialization: technology traps. Ask yourself this: How many times, over and over, has the world witnessed cultures that rise up, prosper, and then completely disappear, often leaving behind little evidence of their breadth of knowledge and sophistication beyond artifacts we can sometimes barely comprehend?

How did the ancient Egyptians or the Maya or Khmer Thais quarry stones for their pyramids and temples so precisely that, even today, you can't slip a playing card between them? This is all the more mysterious because none of these cultures had access to the advanced metallurgy that would have allowed them to cut stone with relative ease. What happened to completely destroy the early farming cultures of Mesopotamia or of the Anasazi people of the American Southwest, where anthropologists can prove that they had farming and societal practices virtually unheard of during the age anywhere else on Earth? And did all of these cultures fail for the same reasons? How is it that dominant human cultures of their respective ages suddenly evaporated or at least backslid, with a change of fortune that often took a wealth of technological knowledge away with it?

To me, the answer is simple. These societies came to rely on one way of doing things to the exclusion of all others. They came to believe in only one technology: one way. And, in doing so, they sometimes pointed themselves toward doom.

Still, among *Homo sapiens,* intelligence and the ability to adapt gave them an obvious way out of the technology trap of overhunting. To survive and grow their numbers, these bands of hunter-gatherers continued pushing outward into the world, hunting new frontiers. And as they did so, they stumbled onto another technology paradigm.

Just as with the Hadzabe of today, among these early hunter-gatherer populations scattered in growing numbers across Africa, Asia, Europe, and the Middle East, while hunting had been the dominant societal force for dozens—if not hundreds—of generations until about 19,000 years ago, it is likely there had always been a "gathering team" to augment the "hunting team." It's often speculated, and widely agreed upon, that these two teams broke down along gender lines, with the tribal males of a certain age doing much of the hunting, and the women and children doing most of the gathering. But as game grew scarce, the push to gather only grew stronger, and soon bands of hunter-gatherers had moved closer to the steadier sources of fruits and grains.

They kept expanding their ranges, too: Working in small hunting and gathering groups, these migrations kept fanning outward across the Earth. In fact, the second oldest DNA found in a living population, that with the fewest markers and the closest to the DNA of the Hadzabe, !Kung, and San Bushmen, was discovered among populations of Aboriginal natives in northern Australia. In fact, science has now found evidence that Australian Aborigines appear to have walked the coasts, proved by ancient DNA markers recently uncovered in India that link the Aboriginal Australians to the Bushmen of the Rift Valley neighborhood. *Homo sapiens* were truly spreading everywhere.

With an ice age locking up water some 25,000 years ago, the first Euro-Asians also began to explore beyond the Bering land bridge, entering North America. Within a few thousand years, they'd migrated far south into Central and South America, as well. Modern humans, *Homo sapiens,* were populating the world.

And then, worldwide and at about the same time, roughly 11,000 years ago, as if flipping a switch, the hunter-gatherers began to cultivate crops and husband livestock, and the transition to the Neolithic, the New Stone Age, was complete. During this era of the modern human's ascent toward today and tomorrow, bands of people blanketed much of the habitable globe within a handful of centuries, and they began to set up housekeeping. It was a hard life, where survival depended on resourcefulness as well as resources. But humankind was succeeding in making the world work for it in a way the planet had never done for any other species.

And *Homo sapiens*, the great communicator and experimenter, was exploiting that advantage any way it could. In a sort of evolutionary Hail Mary pass, Neolithic humans were gaining a foothold everywhere—from Asia and the isolated Channel Islands off the California coast to Australia and the jungles of South America, searching to find what regular sustenance could be wrung from the earth. Within a few thousand years, what had been an exclusively mobile hunting-and-gathering culture spread to virtually all corners of the habitable globe, slowly becoming a decidedly more sedentary species. Neolithic populations took up digging for roots, living in tidal estuaries to scratch the muddy banks for food, setting up in the bogs of Europe to eat whatever fruits and berries they could scrounge, and (perhaps most popularly) establishing outposts near fields of wild grasses and grains that could help to sustain them.

From this shift that emphasized gathering more than hunting it was a short, always-hungry leap for a few of the more intuitive tribal members who came to realize that, when kernels of grain are dropped on the earth, some of them would sprout. Such realizations probably came gradually. Yet, judging by the rise of agricultural leanings during the early Neolithic, farming seems to have arrived on every continent at about the same time.

At a place called Monte Verde in Chile, for example, the archaeological record shows that a permanent village had been erected at least 12,500 years ago, complete with rectangular huts made of wood and animal hides. In the middens there the peelings of ancient potatoes and the husks of maize and other grains were found alongside the charred and blade- and tooth-scratched bones of both small game and the soon-to-be-extinct mastodon.

In other places, such as the Levant, in the Middle East, archaeological evidence shows that populations set up near fields of different grains that matured successively across the seasons, and they ate olives that grew throughout at least three seasons of the year. Around the Mediterranean, wheat and barley, as well as rye, olives, and grapes were beginning to be harvested. In Africa, groundnuts, watermelon, yams, and sorghum were harvested in increasing amounts. At the same time, millet, rice, cucumbers, yams, soybeans, and coconuts were slowly becoming the gathered and then cultivated staples of Asia. In the Americas, maize, squash, beans, potatoes, peanuts, peppers, and pineapples began to be the diet.

Though it's estimated there were perhaps 300,000 to one million *Homo sapiens* scattered around the world between the demise of the Neanderthals and the rise of the first provable remnants of modern human settlements, the one thing that's obvious is that the grains, fruits, and vegetables actively tended by these newly anchored cultures were sustaining a growing number of people. And, somehow, viewed from a station 11,000 years later, though our ancestors were still a few thousand years from becoming true cultivators, the gatherers were not only providing increasingly sustainable sources of food for themselves, they were, through either accident or luck, also starting to cultivate the right kinds of food.

How do we know this? Because the same staple crops that once fed between 300,000 and a million *Homo sapiens* during the Neolithic age today feed roughly 6.5 billion people.

But the life of the Neolithic gatherer-hunter, and the human desire for protein from meat, wasn't crushed out just yet. Though technology had begun to doom hunting and gathering as an easy way of life, as Julius's modern Hadzabe and the Arctic Inuits and Australian Aborigines clearly prove, in some environments hunting and gathering remains a viable approach to sustainability for small and mobile groups that protect their resources. Still, as the human population grew and stocks of easy game became more difficult to find, the Neolithic hunters were smart enough to shift their tactics a little in order to keep meat on the menu.

In the Neolithic, the early cultivation of plants was complemented by new styles of animal exploitation. About the same time humans realized that keeping proximity to regular sources of fruit and grains was a good idea, they also began domesticating formerly wild animals. Perhaps they trapped the first specimens, keeping them alive and breeding them (and coming to drink their milk) until individual animals were needed for food.

In Africa and the Middle East, goats, sheep, camels, and wild asses were being domesticated. In Europe, the emphasis was on cattle, hogs, and geese. In Asia, the Bactrian camel, yak, zebu, chicken, pig, and water buffalo were tamed for work and food. In the New World, at roughly the same time, though fewer varieties of animals were raised, the goal was the same. People were learning to stockpile still-living animals for later food sources, with the archaeological record in Central and South America showing that domesticated species included llamas, alpacas, and turkeys.

So once again, at roughly the same time all across the planet, modern humans were not only starting to take cereals, fruits, and vegetables in recognized patterns on the landscape, they were beginning to maintain domesticated animals, as well.

Though they were still hunters, *Homo sapiens* had greater intelligence than any species that had ever lived. At this point, humans figured out ways to keep their need for animal protein not just met, but slaked. Domesticating livestock and cultivating crops were the first truly liberating steps toward a notion of taking the world as it existed and shaping it for personal needs.

The rise of modern civilization, with its towns and cities and its increasingly elaborate scaffolds of knowledge that stretched farther and father away from simple subsistence, would not trail far behind.

Finally, late in the day, it's time to visit Baalbek's famed Roman ruins. Encircled by tall, ancient limestone walls and spread across perhaps 150 acres, the largest Roman religious complex on Earth occupies a bench-like hilltop between the newer town and the sprawling agricultural valley to its west. And Baalbek's ruins aren't simply the largest Roman religious ruins on Earth or the biggest tourist site in Lebanon . . . they're stunning.

By the time I get to the entrance, it's late. As I pay about eight dollars for my ticket, I feel that the late afternoon heat and the blue sky overhead have left this place eternally in summer. Even before I get to the ruins themselves, troops of lizards scramble across the rock walls alongside small blue flowers with yellow interiors that bob on the ends of long green stalks, their roots and leaves squeezing up through the cracks of the paving stones. Overhead, vultures wheel in the sky on late afternoon thermals. And then, as I walk a gravel path toward the site, moving around the edge of an old, stone maintenance building—straight ahead—the thing looms, impossibly huge. Losing my coordination and stubbing my toe on the rock path, I literally stumble into Baalbek's ancient history, coming first upon the Great Courtyard.

A massive, 450-foot by 370-foot enclosure with four elegant porticoes, semicircular exedras, and statues in wall niches, the Great Courtyard is also sometimes called the Courtyard of the Sacrifices. From today's perspective, its huge entranceway, smaller doorways and window apertures, two central altars, storerooms, and underground passages are comparable in size and complexity only to a modern football stadium. The late afternoon sun streams in as I stand in the center of it, having the place to myself. It's almost impossible to believe that, at one time, a wooden roof shaded the entire structure.

The most amazing feature of the courtyard is its pillars. They aren't constructed of the local sandstone that makes up the rest of the ruins. Instead, they were carved from gigantic liths of red granite; monsters brought to this place from the location of their quarrying at Aswan, in central Egypt. Once removed from the earth, they were floated down the Nile on barges, and shipped north across the Mediterranean to the shores of Lebanon, where they were unloaded and transported all these miles inland to be carved and raised as pillars.

In the stillness of early evening, I turn slowly in a circle, taking in the Great Courtyard beneath a cloudless scrim of sky that's gaining an electric glow from the sunset. On the ground, in every direction, above the courtyard's shattered walls, all of which are carved with entablature relief images of Tritons, Medusas, and Nereids, stone walls and shattered stones lie in testimony to the ego and desires of the human past. Today, there are destinationless passages and stairways leading three stories into the sky, and crumbled pillars lying on the ground in circular sections, never to stand upright again.

Birds tweet in the enveloping darkness. The delicate blue flowers stand tall on their green stalks. Grass and weeds grow in clumps near the altar areas. Each of the stones that make up the walls of the Great Courtyard is the size of an automobile. Each must weight several tons. Who got them here and placed them atop one another? How did this all happen? How can we make sense of it all? Perhaps the most important question is how did this place rise up as the largest Roman religious structure on Earth . . . and end up like this?

Just ahead, bathed in the sunset's glow, is Baalbek's Temple of Jupiter: the single largest Roman temple ruin on Earth.

As a reminder of its size, six of the world's tallest columns, each of them a towering 22 meters, or more than 72 feet, still stand above its crumbled peristyle hall, an open-to-the-sky floor paved with perfectly carved blocks of limestone that extends more than 240 feet in length . . . and spans an area almost as wide. Though today the structure around it has been reduced to rubble by time and almost 2,000 years of earthquakes, the building was once capped by a roof supported by 54 columns precisely the height of the six that remain standing today. The makers of this temple had column-building figured out to precision tolerances.

I walk through the temple, trying to conjure daily life here: I imagine women and men walking, talking, worshipping, and gathering along its porticoes to discuss critical events of the day. More birds move through the air. The late afternoon light has now gone warm and gold; the temple's perfectly laid floor is smooth.

From the far side I can see across the valley. What's most amazing is that, standing where I am, I'm in the treetops. I walk to the ruin's western edge and look down. The westward-facing side drops off like a cliff.

When I look over the wall's sheer drop, it's obvious that the Roman builders put this place on a foundation of huge flat plinths to keep everything steady. I'm standing atop a maybe 60-foot cliff of square, quarried stones brought from somewhere else to here, then laid atop one another to form that foundation. According to my guidebook, each of these stones is 70 feet long, 12 feet high, and weighs 1,000 tons. And they're stacked beneath this place like children's blocks. The builders of this place intended it to endure.

After a few minutes spent studying the fields of the Bekaa Valley, where rivers run into the taupe brown desert earth like veins of lifeblood, I turn back to face the temple. On the opposite side is a smaller acropolis, the Temple of Bacchus, much of which is still standing.

This is a tiny structure by local standards, though it is far larger than the Parthenon in Athens. As I walk toward it, one of the things that grows obvious is that its architects were conscious of its status as historical envoy.

With nods to the civilizations that preceded Rome, the temple includes two square, Egyptian-style columns at its entrance, more than a dozen fluted Greek columns flanking its interior walls, and an encircling exterior colonnade of 42 rounded Roman columns supporting the roof. Looking inside from the grand staircase at the temple's front, I can see fragments of the ceiling, decorated with stupendous, painted bas-reliefs of the Roman gods: Winged Victory, Mars in armor, Vulcan with his hammer, Bacchus and his grapes, Diana wielding her bow, and—given the crops the Romans once grew here—a very prominent Ceres carrying her sheaf of wheat. Because of this legendary bounty, Baalbek and the Bekaa Valley are still referred to as the breadbasket of the Roman Empire.

And really, like the city and the valley, this temple, too, is amazing, impossible. And yet there's something more, a certain fragility. From where I stand, staring, the temple seems on the verge of complete collapse. The huge stones of the ceilings tilt downward and slightly inward. Some of the columns have collapsed; yet somehow they are still upright, supported by the temple walls. In the sunset, the place feels like human culture colliding with nature's rebuke: man creating something to stand for the ages as time, weather, and entropy slowly claw at its existence.

Small, fast-moving swallows flit through the air, devouring insects that have emerged from the grass with the coming of evening. The sunset air swirls with them.

For a dozen or more long minutes, I stand and stare at the building, its bas-relief carvings, and the birds soaring in the sky above it all. This place is beautiful and solid, delicate . . . and quiet. I'm so caught up I barely feel the tap on my shoulder. It's a security guard; in his right hand, he's carrying a large silver flashlight.

"Excuse me, sir," he says in English. "But it's time to go. The ruins have been closed for almost an hour. How did you get in here?"

I dig in my jeans pocket and extract my ticket, showing it to the guard. "I just came, and have been looking around," I say. "This place is unbelievable."

"Yes, yes," he says. "But now it is time to go."

Back at the hotel, Maya is waiting, concerned. "Where have you been?" she asks as I walk through the gate. "I've been worried."

"I was, uh, out . . . looking around," I say. "Sorry. Didn't realize it would take so long."

A wooden table has been set for dinner in the hotel's courtyard, beneath a spreading tree, in the gravel. There's a tablecloth and plates, silverware; candles burn, and a glass pitcher of water and open bottle of wine are there, too. Off to the right is a small reflecting pool. As we're the only guests at the hotel, I assume the table has been set for us.

In my absence, Maya has been busy, as well. Near the table, a well-dressed man with a mustache, probably in his 50s, sits on a stone wall. He's drinking a cup of tea. The man's name, Maya informs me, is Hassan Abbas Nasrallah, and he's the city's historian. Mr. Nasrallah is all business. After we've been introduced and quickly shake hands, he settles back against the stones of the wall, and launches into a fast-moving monologue, describing Baalbek's past.

"There's a lot of mythology in Baalbek's history," he says. "Many historians say Baalbek was built not by man but by the gods. . . . Even Herodotus says this is the mythology. The Phoenicians said this in their mythology. And while it doesn't make a lot of sense rationally, perhaps the myths are right. Perhaps the city was built by gods, not human beings."

"In fact," Mr. Nasrallah goes on, "despite the fact the temples and ruins of the ancient city have been crawled over by archaeologists, no tool used to actually build the city has ever been found."

It sounds like a bunch of hooey to me, but Mr. Nasrallah still isn't done. He sips his tea. "Even the site of this very hotel," he says, "this very place we sit now, was once a Roman amphitheater. It's under this hotel. Look around. . . ."

I do. And I can see exactly what he's saying. With its tiered encircling walls obscuring the outside, each level about the width of a theater seat, this place does look like a filled-in former Roman forum.

"Scratch the surface anywhere in Baalbek, and you'll find not only history but mythology and mystery," Mr. Nasrallah says, not pausing for me to keep up. "The city has constantly changed over time. During the Roman period, for instance, the road just outside this gate was bigger: it was 28 meters [or 92 feet] wide, with pedestrian areas and temples on both sides of it. It was a road for large armies and transport and commerce. When the Trojans went to invade Persia, this road carried them into the city, then directed them onward, and along its path is where they stopped. The city—and the valley—drew them here. Baalbek has always been an open city, too: open to the gods, astrology, and black magic. Baalbek was good for all religions, all kings, all men. Here the earth blossoms as art . . . and the stones speak poetry."

Mr. Nasrallah has his momentum up. He's rolling. "In Baalbek, under every stone, there is something of historic and archaeological value, but nobody wants to take the time to dig. There's something under every single rock, but nobody wants to look."

"That's true for a lot of places," I say. "That's true for the world."

In the falling night, Mr. Nasrallah sips his tea some more. As he does, a server comes outside. She is carrying a tray filled with small plates—one of hummus, one of pita bread, and a third mounded with green salad. Other, larger plates hold grilled mutton chops and overflow with piles of golden-yellow French fries.

"We are ready to eat?" the server asks. "It is getting late. The kitchen wants to close."

We ask Mr. Nasrallah if he'd like to join us. "No, no," he says. "I must go home. My own meal is ready tonight." He sets down his tea cup and stands to go.

"No, really," I say. "Please join us. I feel terrible for keeping you waiting."

"No," Mr. Nasrallah says. "I must go. But as you explore the city over the next days, remember this. Natural characteristics are what singled Baalbek out. Water. Wealth. Fertile soil. Crops. The city is protected by mountains on three sides. It is almost unreachable. Mostly, it is these factors that gave Baalbek its future. These natural gifts couldn't be replicated anywhere else on Earth. Not in Turkey. Not Israel. Not Rome or France or North Africa. Nowhere on Earth could have the same gifts as this place."

In recent years, just like the Temple of Bacchus that today seems on the verge of crumbling, the natural gifts of this city and valley have grown tired, embattled, and even a little tarnished. After a wonderful dinner, a good night's sleep, and a morning spent exploring the town, I meet up with Maya after lunch, and we head onto the street to find a willing taxi driver; we then head off into the agricultural lands around

Baalbek, hoping to more closely examine the breadbasket of the Roman Empire.

Driving out of town to the south and east along Rue Abdel Halim Hajar, we're soon into the sun-scoured fields of the Bekaa Valley, the city falling away behind us. But unlike yesterday, when my attention focused on the city in the distance, today my interest is directed more to the land around it. And, frankly, the place is a mess. Away from the rushing blue waters of the aqueducts of the mountainside, the irrigation canals in the valley are a murky and slow-moving brown. Tattered white plastic bags printed in black or white or blue cling to tree trunks and branches or nest in woody shrubs.

Where there are signs of habitation, small huts that have been assembled from wood, faded cloth, and plastic stand near small herds of goats and sheep; the animals nibble on low piles of garbage and rotting vegetables, looking bored. Once in a while, the car has to overtake donkeys that pull two-wheeled wooden carts loaded with forage greens bound together in sheaves. The cart drivers sit atop their loads, while holding the reins and smacking each donkey's haunches with a light, cane-like stick.

Out here amid the fields, the earth of the Bekaa Valley is a pale, gray-brown. Despite all the water, the soil seems dry and not bound together by much in the way of organic or biodynamic particles. It seems, in fact, more like that broken mix of clay, sand, and silt you'd find on a baseball diamond's base paths. The earth here seems tilled and broken and sapped of vitality.

We drive farther along, cruising a network of single-lane farm roads. It's a little disorienting. Unlike yesterday, when the experience of riding into the Bekaa Valley and Baalbek seemed like the approach to an enchanted spot, today the place feels out of focus and exhausted. In some of the fields, long strings of spindly tomato plants are growing. In others, sheets of shiny black plastic have been spread across the soil. This material attracts the hot sunlight and heats the earth so severely that no weeds can survive beneath it. In other fields, rows of green forage are being cultivated. And everywhere are the blowing, fluttering pennants of shredded plastic bags. How is this the same place we saw yesterday?

Finally, I see a good place to stop. A row of tall trees planted along the single track will provide some shade for the driver. In the distance, a windblown hut of scrap wood, blue canvas, and plastic sheeting stands in a field. Next to the shelter, a man wearing blue peasant robes and a loose head wrap of grayish white muslin is seated on a woven plastic rug; children swirl and play around him, amid a small herd of sheep and goats. I ask the driver to stop.

I step from the car, and, with Maya trailing behind, wander toward the bereft-looking camp. *"Salaam Alekum,"* I say to the man, who is still seated quite a distance away. "Peace be with you."

The man stands and raises his right hand. *"Alekum as Salaam,"* he says back. "And peace be with you." He steps into his leather sandals, which sit in the soil just off the carpet, and begins walking from his ragged camp out to meet us.

With Maya as interpreter, we begin to talk. The man seems circumspect, but friendly.

"What is your name?" I ask. I pull out my notebook.

"Mohamed."

"Do you have a last name?"

He glances at my notebook. "My name is only Mohamed," he says. "Only Mohamed."

"Do you farm this field . . . this area?'

"Yes," he says. "I have ten children. I live here. I farm here."

"How does this go?"

In the bright, hot sunshine, Mohamed slaps at the air with his hand. "It does not go so very well any more," he says.

The Stone Age, it has been noted in other places, did not end because humans ran out of stones. It ended when *Homo sapiens* made a huge Neolithic-age leap forward in thinking and technology. And much of that leap was facilitated by one change in human practice. Slowly, but with increasing volume and efficiency, modern humans switched from hunting and gathering to a committed life of farming.

Beginning about 11,000 years ago, Neolithic humans were stumbling into the earliest stages of a new, more agrarian way of life. Still, the route remained slow going, with heavy work and the threat of starvation always nearby. Nobody would be jetting off to vacation at Cabo San Lucas or Ko Samui any time soon. But over the next thousands of years, the archaeological record shows that humans learned how to grow cereals and breed animals in incrementally increasing numbers.

It's important to stress that, just as modern farming is fraught with challenges, early efforts had their ups and downs.

Still, even in its early precariousness, agriculture was, to many people, far preferable and more energy-efficient than the life of hunting and gathering. Thus began a positive feedback loop that, in terms of its magnitude, compares with nothing else in the cultural history of *Homo sapiens.*

Farming changed the history of humankind: With the human cultivation of crops, the crops cultivated the humans as well.

"The establishment of farming suddenly meant humans had time for other things," says Rae Lesser Blumberg, a professor of sociology at the University of Virginia, who has studied ancient and modern farming and its effects on society for five decades and in more than 40 countries. "And it didn't take long, only a few thousand years, for what was once a subsistence culture to become a solidified one that had a centralized village or city, with stratification and division of labor inside that society. With some exceptions and notable 'buts' along the way, farming set up what human society has become. Not all cultures progressed at the same speed; many progressed at very different speeds. But in the story of what modern humankind is, farming is a huge component. Again, with some exceptions, the idea of farming might well be the differentiating component between ancient and modern humans."

As mentioned, by 11,000 years ago Monte Verde in Chile was established, its society thriving as people dined on grilled animals and potatoes from the local harvest. Not long after that, in the lowland areas of the Middle East's Levant, communities were sprouting up, many of them permanent enough that the land soon became barnacled with small mud-brick or stone buildings.

At localities like Tel es-Sultan in Jericho, today located within the West Bank Palestinian Territories, or at Çatal Hüyük in central Turkey—places well watered thanks to sometimes snowy mountainsides and springs nearby—ancient towns took root. Tel es-Sultan, said to be among the oldest continuous settlements on Earth, is so ancient that excavations show its earliest structures were devoid of clay containers, rendering it a rare "pre-pottery" Neolithic site. At the deepest levels of the *tel*—the Arabic word for a mound of rubble left from previous civilizations—only stone bowls and stone tools have been found.

But even if pottery had yet to make its debut at Tel es-Sultan, the archaeological evidence shows that by 10,000 to 12,000 years ago, the humans living there had made an active decision to stay put and farm. The roughly ten-acre site is contained within a protective wall of piled stones, which suggests, among other things, a social organization that was sophisticated enough to work collectively for protection from what may have been aggressive outsiders or dangerous floodwaters.

Excavations at Tel es-Sultan also show that a formalized recognition and honoring of the dead was firmly in place. Human skulls, covered with natural-looking and "skin-like" layers of clay and eyes of white cowrie shells have been discovered; macabre suggestions that people wanted to keep the memory of honored citizens and family members alive.

At Çatal Hüyük, which appears to have been constructed at about the same time, a honeycomb-like network of tightly settled houses stretches across 32 acres. And although no protective wall encircles this site, each of the houses there could only be entered through the roof, an adaptation that provided its own form of security. There were no streets: The linked rooftops were the streets, again implying that protection from possibly violent outsiders was required, as was a means to escape the stifling on-the-ground heat of summers in central Turkey.

It's strange . . . or maybe not so strange. But with the early rise of farming—of people becoming established in one location for long periods of time—personal security from marauding outsiders suddenly appears as a primary human goal. Even in the earliest evidence of settled Neolithic culture, human-on-human aggression was becoming an ongoing concern. Which also implies an established tribalism among those working to protect themselves.

"As humans learned how to keep livestock," says Rae Blumberg, "they were also learning the first few things about organized warfare. Earlier than this, in the age of hunting and gathering, we see very little evidence of warlike behavior. I'm not saying there weren't disagreements between bands of hunter-gatherers. But with the rise of farming, we suddenly find settlements inside walls. Warfare leaves tracks; archaeological evidence. And the rise of farming and settled societies required protective measures for those who wanted to avoid aggression. I mean, it's always been easier and more efficient to take another village's livestock than to grow your own. So there it is: the basis for everything from wars to corporate takeovers that comes after. *Homo sapiens* have been doing this for a long time. And the archaeological proof is there, in those first stone walls."

With those first barriers of hand-laid stones, an obvious new world of Neolithic life is exposed. Among this new form of human society, each stone that was intentionally set into that wall carried with it the same point. Life had been distilled down to *us* and *them*.

Still, Çatal Hüyük is also archaeological proof that an early human farming society was progressing nicely. Cooking was being done indoors, on mud hearths made specifically for that purpose. The structures, with their mud-brick walls covered with white plaster, were cool in summer and warmed by the fires in winter. Cattle heads, covered with plaster and hung on the interior and exterior walls, were used for decoration. Flax was being cultivated, and early textiles were being created from that flax.

At the Museum of Anatolian Civilization in Ankara, scraps and fragments of beautiful—if rough woven—cloth excavated from the ruins of Çatal Hüyük sit inside Plexiglas display boxes, cloth that may be 10,000 years old. Also found at Çatal Hüyük are tools made of volcanic obsidian glass, which had to be imported, implying either wide-ranging travel by members of the community or trade with outsiders. The museum also houses sculptures from the site. In some of the smaller outer rooms of the structures at Çatal Hüyük, likely storage areas for the grain and foods grown nearby, small and sophisticated terra-cotta sculptures of seated human forms have been found that are believed to be deity figurines, placed in the room alongside the food as a way to protect it.

And perhaps most important: The archaeological record proves that the farming people of Çatal Hüyük had enough free time for things beyond survival. Aside from the plaster covering and decorative cattle skulls, the walls are decorated with vivid murals and frescoes with skillful representations of wild cattle and deer and people. Painted in pigments that had to have either been invented by the inhabitants or traded for, these murals exist on both the internal and external walls of houses; showing that the locals had a growing understanding of symbology to make visual contact with others through signage, not to mention enough leisure time and material wealth to either create these symbols themselves or hire someone else to do it. Unlike the rough representational drawings on the inside of caves of France and northern Spain, which earlier *Homo sapiens* had probably been doing for their own entertainment, these people were sending messages out into the world.

A recognizable human language of ideas and symbols was rising, though the earliest written words were still thousands of years in the future.

Yet alongside the successful ascent of human farming, there came a companion fear. With established towns and annual cereal planting and harvesting—not to mention the development of some domesticated animals—there grew a sense of entrapment. What happens should things go bad?

Twelve thousand to 10,000 years ago, for growing numbers of farming people, a new plan for life was forming. They had property, not to mention probably larger and more extended families than had existed among hunter-gatherer bands. But this made them far less mobile. Suddenly, they had to continue farming successfully or face starvation.

"It also probably meant, though, that older members of these groups were suddenly in higher regard," says Rae Lesser Blumberg. "With no written language, and a farming technology that still was young, meaning they had some hard years, the old [people] in any village or society were the institutional memory. They were the ones who knew what happened the last time there was a drought; the last time the crops didn't grow the way they had in the years before. Back then, the elderly were probably greatly prized for being a link to survival. A settled life of farming was changing society and the way it conducted itself."

Though prospects for humanity were growing, and humans themselves were discovering more free time and ease, in many parts of the world humans had made a conscious decision to step away from hunting and gathering. And though hunting and gathering had its own problems, its societies had survived drought and blight and fire for tens of thousands of years in a way settled farming communities had not. Still, these Neolithic agriculturalists, even with the newfound stability and ease farming provided, were about to learn of new problems. These would be ideas like overgrazing and soil depletion and outsider aggression. And these new threats were coming fast.

---

Back in Mohamed's fields of the Bekaa Valley, the ideas of overgrazing and soil depletion—not to mention exterior assault—have become far bigger and less abstract ideas. According to Mohamed, there are lots of reasons farming in the Bekaa Valley has grown to be a difficult task in recent years. "And many of these problems are in the soil itself," he says. "The ground has grown tired. This is due to many things."

Primary among these influences, according to Mohamed, is this: Like everything else in the Middle East, in recent years the quality of an agricultural life in the Bekaa Valley has been overlaid by religion. Though the valley is said to contain 40 percent of all of Lebanon's arable land, the more fertile, lower-altitude, southern portions receive most of the interest and fertilizer, producing rich yields of corn, wheat, cotton, and vegetables. The southern region is also land that, not incidentally, is largely owned and farmed by Roman Catholics and Maronite and Eastern Orthodox Christians: non-Muslims who possess more influence with the national politicians.

And while the northern end of the valley does still have irrigation and water-distribution systems that date all the way back to Roman times, these aqueducts are subject to droughts and seasonal fluctuations in water levels. Down in the southern end of the valley, since 1957, irrigation has come about through the Litani Hydroelectric Project, an involved series of dams and canals underwritten by the government.

"Here in the north of the valley," he says, "the land is less fertile due to its location. Soil nutrients flow downhill, with the water. And here, we farmers are mostly Sunni [Muslim]."

*A Lebanese Arab man outside Baalbek*

Also, when the most recent shelling between Lebanon and Israel fired up again in 2006 as retaliation for rocket attacks, many of the farm laborers, who were Syrian, left the valley and went home. Still, the problem is far older than 2006, though the Israeli shelling that time did target and destroy 20 percent of the habitable structures in the Bekaa Valley while rendering many of the crops that year unharvestable and the fields dangerous.

Even now, unexploded ordnance is still sometimes found in the fields. Thirty-three people were killed during the 2006 assaults. Also, because of the less fertile soil in the northern end of the valley and the lack of influence with the central government, each year more of the cropland in the valley's northern end is being grazed by livestock.

"Growing numbers of farmers are leaving, or they are becoming nomads," Mohamed says, his hands flipping up, palms remaining pointed downward. "They just graze animals on fields that used to produce vegetables. Animal forage is all that can be done there successfully anymore. Or they have started growing drugs, marijuana and hashish, to support themselves. This provides good money."

It's true. According to Oxfam International, of the 195,000 farmers in the region, 75 percent of them own one hectare (or 2.47 acres) or less. And most of them, in 2006, showed losses in crops averaging about $35,000. Using Mohamed as an example, a simple and low-tech subsistence farmer with ten children to feed, it's easy to see how those kinds of losses—even in a single growing season—could have disastrous implications. The economic and market pressures could force him into new and more cost-efficient lines of crop cultivation.

"Do you grow drugs?" I ask.

"Oh, no . . . no . . . " Mohamed says. He waves his hands in the air. He smiles imploringly. "No, I do not grow any drugs here. But others do, since it can make them more money off land that is now tired."

He pauses for a long time, looking away. "And in truth," he says, "the farmers are growing tired, too. There is so much today that distracts them, makes them worried. For many, farming drugs is easier and more lucrative work."

Mohamed invites us back to his house, the windowless, 10-by-12-foot shanty of boards, plywood, and blue canvas and plastic tarps that's toward the middle of one of his fields. A black stovepipe emerges from the house's slightly sloped roof. As we begin walking across the soil, the earth is loose and gray and plowed into low, rough furrows. Mohamed points out small green sprays of vine and leaf every foot or two.

"These are potatoes," he says. "Right now on this field, potatoes are the best crop I can grow. I need fertilizer . . . and then to let my fields rest. Potatoes do not make a farmer a lot of money, but they can grow in soil that is depleted. I have ten children . . . and a wife, and animals, and . . . potatoes. Other things in other fields, too, I know this. God is great. But on many days, like this ground, I am exhausted and wondering about the future."

As we approach the house, Mohamed asks that we sit on a carpet outside. Beneath a blue sky, with birds coursing between lines of trees overhead and the city on the hill behind us in the distance, it's amazingly peaceful. The sun is still strong, but not overpowering. Stretching away in every direction are cultivated fields. Inside the house, a cooking fire must be going; smoke drifts from the stovepipe up into the sky. "I will get you coffee," Mohamed says, stepping through the doorway, its canvas door folded up onto the plywood and canvas roof.

"No, no," Maya and I say. "That is not necessary."

Mohamed shrugs and smiles. "Of course it is not necessary . . . but you are visitors. You are my guests. I have invited you."

Self-consciously, we continue standing. Inside, crockery and silverware clink. A few minutes later, Mohamed comes out, bearing a small plastic tray with tiny coffee cups on it. Lebanese coffee, I have discovered over the last few days, is very strong. A small cup is enough. Two cups leave you almost jittery.

Mohamed kneels and sets the tray on the carpet between us. He removes his sandals. Then he sits. "Obviously, it is *sadah,*" he says, meaning unsweetened. "But, if you would like, I believe we have sugar for your cups."

No, we're fine. We kneel onto the carpet, shoes still on, the bottoms of our shoes pointed away from everyone to avoid the Islamic slight of pointing the sole of your foot toward another person.

"You would like food? Some small fruit? Bread? Something?"

"No, coffee is enough."

We sip the coffee, which is screamingly hot. Mohamed's children watch us from inside his shelter's dark doorway—and also from around the edges of its exterior. They're aged from about 3 years old to perhaps 15. Goats and donkeys continue to mill around, too, though some have spread out a bit and are browsing the vegetation along the field's edges. When one of these animals gets near a potato plant and ducks its head to begin eating, the elder children chase it away, some of them tossing pebbles at it. The coffee has cooled a bit. We can sip it now. As we drink, Mohamed looks around, watching everything.

"How many fields do you have?" I ask.

"This one, and a few others beyond the trees. Plus on the other side of the road."

"But you have enough?"

"Yes, yes, God willing," he says. "We are near to the irrigation, so there is good water. We are still able to grow things and don't have to just graze the animals, though, as I have said, we need fertilizer to help us grow more. But yes, we have enough. Every year is a lot of work, but we always have had enough in the past. And friends to help."

"Do you live in this place all year?"

"No," Mohamed says. "In winter, we move to Baalbek, with my wife's family. They live up on the mountainside, a little east of town; above it. We pen the animals near here in winter, and I come most every day and feed them. In the spring, I return to cultivating the fields during the day, but go home at night. Because of the altitude, nights can be cold here, even in summer. Slowly, though, I begin spending more and more time here. I spend a few nights. I get the place ready for summer. Then, as school ends, we move here as a family: my wife and children and me. The children play in the fields and irrigation streams. The older ones help with farming. Friends from other farms nearby can be asked to help if I need more adults. I just walk over and ask. It seems primitive to you probably, I know, but this life has many good parts."

"With ten children and your wife, where does everyone sleep?" I ask. "This house does not seem big enough."

"We all sleep here," Mohamed says. He points at the ground, then turns his index finger at the house. "And some nights, our family from the town comes out, too. We eat a good meal, and then roll out the carpets and cushions and blankets that are now rolled up inside the house. We roll them up and store them every morning. We sleep inside and out. As I say, even in summer sometimes it can be cold at night at this altitude. But this is our life for much of the year. We farm, we eat a nice meal; we see the stars at night in the sky. Then we roll out the carpets and blankets and sleep, all of us together. A family."

We pause and finish the coffee. Mohamed is right. It's pleasant here.

"Do you not worry about the future?"

"Of course I worry for the future," he says. "But it is the future. All I can do is make today as good as possible with what I have, then hope that goes forward in the future. The future starts every day."

"Do you not worry for your children?"

Mohamed cocks his head toward me a bit. He looks me directly in the eyes. "Do you have children?" he asks.

"Yes," I say. "My wife and I have two children: a boy and a girl."

He smiles. "Do you worry about *your* children?"

I smile and shrug. "Of course," I say. "I asked a silly question. Everyone worries about their children."

Mohamed smiles and tilts his coffee cup, examining its contents. He takes a sip. "I think probably most all parents who ever lived want more for their children than they have for themselves. Do you not think, since the start of time, this has ever been different?"

In Baalbek, late every afternoon, I walk over to the ruins and wander, often having the place all to myself, complete with its history, birds, lizards, and blooming wildflowers. Visiting the different temples and halls over several days, I notice that, while Rome had erected these buildings across almost four centuries, starting more than 2,000 years ago, much of the decorative work was left undone. In some places, entablatures or walls are only half-carved. Obviously, these buildings were first built, then decorated.

And, I keep reminding myself, the Romans weren't simply raising temples in Baalbek, they were farming the valley and creating the aqueducts, too. So each evening, I walk to the back of the Temple of Jupiter and overlook the braiding stone aqueducts filled with gray-green snowmelt; torrents whose weighty, downhill-rushing pressure once powered the city's now dry fountains, baths, and reflecting pools as well, smaller monuments that still stand among the ruins and throughout the surrounding town.

On my second night roaming the ruins, at the hour of sunset, I wander out, not wanting to be busted by security two nights in a row. At the entrance I meet two of the head guides, Mohamed Wehbe and Khalid Abbass. They are closing up for the night. If I'm related to any Lebanese Arabs in Baalbek, I hope it's these two guys. The next night as I'm exiting the ruins, they wave me over toward the ticket office, a low white building of plastered-over cinderblock, surrounded by thick lawns.

"You have been here now two evenings in a row," says the guide whose name is Mohamed. He is the taller and darker-complexioned of the two. "Come over, have some coffee at the end of the day. We've already started making it. It is almost ready. This is something we often do, to end the working part of a day."

I step toward them, and as I explain what I'm doing in Baalbek they pull out tiny white cups and a plastic tray. The coffee, which has been made in a tiny steel boiler, is poured. It is, once again, thick and looks ferociously hot. "Come," Khalid says, "let's sit outside." We step from beneath the front door and overhanging roof of the ticket office. Around us, blocks of stone quarried by ancient Rome rise up as walls. Sunset colors the sky with a thousand blues and reds and yellows.

Mohamed and Khalid sit on chairs. I take my tiny coffee cup off the tray and—with only two chairs available outside—am attracted by a carpet of grass across a narrow sidewalk from the chairs. Taking a step away, I begin to sit on the grass.

"Please," says Khalid, "take my chair. I am sorry. Mohamed and I do this every night, in these same two chairs. So if a third person is here, we don't think of it. Please, take my chair."

"No, really," I say, settling onto the grass. "It's fine." The grass is cool and very soft feeling. It's nice. Looking around at all the quarried stone, I marvel at it. "I can't believe the way the Romans built this place," I say. "They did it so quickly, too. It seems impossible."

Khalid and Mohamed smile. "You have probably heard about gods having created these structures in ancient Baalbek," Khalid says. "Also you may have heard that no tools have ever been found . . . and the like. But that is not true. Men made this place. They used tools. Tools have been found. But by the time they began making these temples, people had also gotten very smart. In fact, a single man could quarry out even the largest of stones; the ones for the foundations that hold the rest of these ruins up."

Khalid and Mohamed explain. From a quarry wall, a stoneworker would measure and score a rock to the size he wanted. Then he'd drill an incision deep enough at the stone's back where it attached to the quarry wall, and into this incision, he'd snugly knock pieces of kiln-dried wood. When the wood was packed tight, he'd saturate it with water. In a few days, the expanding wood split the stone precisely.

"It's true," Mohamed says. "If you want, come by tomorrow a little earlier than now, and I'll take you to the quarry site. It is only a few kilometers from here. You can see how they did it. I'll show you a huge block of stone that is still there, waiting,"

Tonight, though, it isn't the mechanics of Greco-Roman construction that Khalid and Mohamed want to teach me. "Because of our history in Baalbek, we are a mix of cultures,"

Mohamed says. "I mean, look at me . . ." Mohamed lifts his right hand and gestures at his black eyes and olive-skinned Mediterranean Arab features. "My wife is colored very much like me. But one of our daughters? She's blue-eyed and blonde. In each of us in Baalbek, we carry many, many cultural and physical influences. So in Baalbek, we understand our mix of cultures, and genetics give us surprises."

Mohamed goes on. "In each of us in Baalbek, the genetics in our bodies are symbols of this place: a crossroads of man. Our religion is equally mixed," he says. "In the Middle East, people argue religion with life and death, but in Baalbek we are peaceful about religion. I am Islamic. My wife and children are Islamic. But once my daughter asked: 'Are we Islamic-Catholic or Islamic-Maronite?' She did not know that Islam and Christianity are different. And anyway, religion means less if you live in a place like this, already as beautiful and peaceful as heaven."

Khalid, in comparison, looks nothing but European. Though he is also Islamic, with his thin face, pale brown hair, long nose, and light coloring, Khalid could be a Catholic native of Milan or Paris. "My father had blue eyes," Khalid says. "Which is why I look European. But I also am Phoenecian, Roman, Assyrian, Babylonian . . . who knows?"

For both Mohamed and Khalid, they say their presence in Baalbek comes down to a single story, one that has little to do with religion or history, but is instead about the point of every day. "It is about King Solomon, who is mentioned often in the Koran and the Bible, and was a great and powerful man," Khalid says. "Well, he was building a temple to himself. And he told the workers, keep decorating the stones, keep inscribing pictures of me. He was old by this time, so he went and sat down on a rock, holding his head up by placing his hands on the top of his cane and resting his chin on top of his hands."

"He died there,' Mohamed says. "But sitting as he was, he was balanced and didn't fall over. And he sat there for weeks and months, until . . . what are they called? Oh yes, insects. These insects chewed his cane and it snapped. Then he fell." Mohamed rocks his body down and to the right, pantomiming a dead King Solomon falling to earth.

Khalid wags a finger in the air. "So the moral of this story and the message of Baalbek are the same," he says. "Life is short, but the work of that life will always be there. Life gets finished, but the work is never finished. The work goes on forever."

We sit for a time, absorbing the message in this beautiful place. In the last of sunset, our coffee is finished, too. Everyone has had a smaller second serving. And then, in the gathering dark, both men are sitting silently and staring at the growing grass. As I observe them watching the grass grow, I realize that with a big bag of books, a hankering for grilled lamb steaks and fried potato dinners, and the luxury of a week to spend purely cooling out, I could pick no better place to enjoy it than kicked-back Baalbek.

I stand, put my coffee cup back on the tray, and thank Khalid and Mohamed for their time. They both stand, too, and shake my hand. In the drawing darkness, we walk the tray back inside and set it on a counter. "I'll wash the cups in the morning," Mohamed says. "It's time to go."

"Come and see us again tomorrow," Mohamed says as he locks up the office. "And if you want, I will take you up and show you the trick of how man made the square rocks for the foundations of this place out of the big rock that is the world." He slips the keys to the building in his pocket. "But now, I have to go home. My wife is waiting. She has dinner."

We begin walking out, with Mohamed and Khalid both heading for their cars near the Baalbek ruins' gate.

"You know my favorite part of this place?" Khalid asks. "It's this. In the winter, sometimes it snows a lot. It can snow a meter or more deep. And then, these ruins are more lovely than at any other time of the year. The layer of snow makes everything more silent and softer looking. The ancient structures and rounded shapes of the columns fallen on the ground under the snow and beneath the sun and the blue sky that often follows a storm. It's really, really beautiful."

Mohamed and Khalid are at their cars now, which sit on a cobblestone street. "This was a good finish to the working day" Khalid says, "As you go home tonight, I hope you have absorbed our message, the message of this place: Enjoy your days, enjoy the gift that is your life. Because life is short, but the work, well, you know. . . ."

At around five the next evening, Maya and I swing by the ticket office at the Baalbek ruins.

"Good to see you," Mohamed says. "Good evening. You came back. I had hoped you would. We will go up to the quarry, so you can see how the foundation stones for this place were made and taken."

He cleans up a few things on his desk, making piles to be addressed in the morning. Then, after he hands off the keys for the office to Khalid, who also says hello, we walk outside, climb into what I recall as Mohamed's brown Toyota, and drive south and east out of town.

The road snakes past our hotel, up the hillside behind it, and, following a few hairpin turns, ends at a gravel parking turn-out near a long stretch of flat hillside. As we arrive, Mohamed explains that, once a big stone was cut from the quarry, it was brought downhill through a simple if involved technology. The ancient engineers selected the stone so it was elevated into the air on one side or another. Then, as they cut the stone free of the earth, they placed the elevated side of the stone onto a series of round wooden poles laid sideways beneath it. That way, they could roll the stone to the temple site by advancing it over the poles, removing the last poles from behind and moving it to the front. It was simple work, if intensive.

"Well, anyway," Mohamed says, "here we are."

What Mohamed wants to show me is nothing short of stunning. Called the Stone of the Pregnant Woman, it is a perfectly squared limestone plinth 13 feet in width, 70 feet in length, and 15 feet in height. Estimated to weigh around 1,000 tons, it is thought to have been an extra foundation stone for the temples that went unused. According to people who've looked into it, there's no crane on Earth today capable of lifting this rock. There was only one way to move it, by pushing it from behind, on long round pieces of wood that are rolled slowly along the ground—the ancient version of ball bearings. Today, the Stone of the Pregnant Woman's top is pitted by almost 2,000 years of weathering and, no doubt, a bit of touristic affection—there are visible pocks and indentations—but the stone is there, waiting.

"They cut this rock with kiln-dried wood?" I ask.

"Well, this is how they separated it from the rock surrounding it, we believe," Mohamed says. "Then they shaped it more with tools."

At the quarry there are two modern buildings: a small, stained-wood tourism office and a restaurant. Both are closed for the night (though a steel rack displaying glossy color brochures about the area's attractions has been left out). I ask Mohamed if I can go and examine the column close up.

"Of course," he says. "But, if you don't mind, I will wait here."

I wander down into the quarry pit, then climb onto the monolith, which is perfectly squared-off . . . and remarkably smooth. As it turns out, there are numerous theories about how ancient man quarried stones like this: for temple foundations in Baalbek, for the pyramids in Egypt, and for the Maya temples and public squares of Tikal in what is today Guatemala. Some theorists believe these ancient cultures developed a technology to "soften" the stone with water or vegetation-based acids, leaving it easier to form. Others think it was purely the result of man-hours and slavery, plus rudimentary metal tools. Other theories subscribe to the idea that these ancient cultures had techniques we haven't yet come to understand, technologies that—according to some, as irrational as it sounds—were brought here by cultures from other planets. Somewhat remarkably, no one has ever been able to prove a single theory conclusively; though, personally, I think the stones were likely cut with a mix of stone-softening soaks and sweaty work by likely undercompensated men using tools. Why should things have been different in the past than they are today?

As I stand atop the Stone of the Pregnant Woman, one thing is clear: At minimum, this enormous carved rock exists, tilting uphill and lifting into the air, awaiting the wooden lengths of "bearings" to be set beneath it for transport.

For a few minutes, I walk up and down the upper surface of this foundation stone. With the exception of some newly formed weathering pits in its top, it's perfectly flat, its corners 90-degrees squared and clean, like the table edge of a linen tablecloth. The Stone of the Pregnant Woman would have been the biggest stone in the entire Baalbek structure. Why do they call it the Stone of the Pregnant Woman, I wonder? Because it's the biggest? Because of what it births? It flashes me back to the huge baobab I visited with Julius: a tree pregnant women climb inside for birth, a tree that holds life-sustaining honeycombs in its high branches.

As I turn to begin climbing down off the megalith, I see Maya standing on the dusty hillside nearby, holding my camera. "You are getting to do something most people can't," she says. "I want to take a picture."

In the sunset, she snaps off a few photos. Then I climb down. Some believe the Stone of the Pregnant Woman was to be one more of the cliff-like foundation plinths beneath the ruins in the valley below; though, in the end, it wasn't needed. By the time Maya and I return to the car, Mohamed has stepped outside, and he's leaning against his car, looking across the valley in the sunset. Nearer to the valley's center, rising in the midst of the ruins, we can all see the six pillars of the Temple of Jupiter towering above the trees and the city.

As we walk back to the car, Mohamed sees Maya carrying my camera. "Will you take a photo of me together with my new friend, Maya?" he asks.

"Of course."

I snap a photo, the Temple of Jupiter pillars clearly visible far in the background. And as we open the car doors and climb in to head back down the mountainside to town, Mohamed pauses for a moment. "You know," he says. "Something Khalid said yesterday has stayed with me since then. Do you remember when he talked about the ruins in the winter? With snow?"

"Yes."

"Well, I have to agree with him, those snowy mornings are some of the best days," Mohamed says. "In winter, we don't get very many visitors anyway, but on days with a heavy snow, we get almost no one coming. Still, we go to work. We make a nice fire in the iron stove in the ticket office, and we wait. We drink coffee. We stay warm inside against the cold. But always, by about eleven o'clock in the morning, I have to go into the ruins and look around. On any day, it's beautiful there, but the snow-covered mornings are often the best. Perhaps this is because you're there by yourself. But to stand in the middle of all that, with a layer of snow covering everything and no footprints anywhere, plus the added silence a blanket of snow brings, well, nothing compares to the beauty of that. The brightness of the snow. The sun. By that time, late in the morning, the snow is also beginning to melt. Water is beginning to flow out of the snow. The world is alive."

Mohamed turns the key and starts his car.

"On those snowy mornings," Mohamed says, "sometimes I cannot help but think just one thing: Every day in the world is something like a gift."

*A Tajik man—another genetically distant relative—with a new family addition in Samarkand, Uzbekistan*

# UZBEKISTAN

**OF ALL THE** destinations on my DNA Heritage Tour, distant Samarkand, in central Uzbekistan, is probably the most unlikely.

"I know, I know," Spencer Wells said when I asked about it. "Central Asia seems far-flung. But your DNA markers spread to that part of the world. And there's no better place to meet people you share DNA ancestry with, tribal Tajiks, than in Samarkand."

So two days after leaving Baalbek I find myself in Samarkand, arguably the Silk Road's most famous traffic circle. Founded in the sixth or seventh century B.C., and known as Marakanda to the classical Greeks, Samarkand can claim to be as old as almost any still going city the world might nominate. But despite its history, double-landlocked inside mountain ranges and deserts and cloistered for almost a century inside the former Soviet Union until a collapsing U.S.S.R. released it in 1991, I doubted Samarkand had much zing to offer me, its prodigal son.

I couldn't have been more wrong. In fact, after finally arriving in Samarkand, home to tribal Tajiks with the genetic markers *M45, M9,* and *M207,* I would come to understand the sentiment that Alexander the Great is said to have expressed upon his arrival in the city back in 329 B.C.: "Everything I have heard about Marakanda is true, except that it is more beautiful than I ever imagined."

Located at the crossroads of ancient India, China, Persia, and Russia, Samarkand has, across human history, often been central to human affairs. In fact, it's only been in recent years that the city has fallen on hard times, having tumbled so far from political and commercial favor that even its airport was closed to commercial flights in the months running up to my visit. To get there these days, you have to journey by car from the national capital of Tashkent, a city whose airport is still functional, though crumbling a little around its poured-concrete edges.

But for me, departing Lebanon, the trek is longer: beginning after a trip back to the Lebanese capital from Baalbek and a last night in Beirut. It's also a journey that turns out to be remarkably non-pretty.

Anyone wanting to make the trip in a single day has to leave Beirut early in the morning, flying by connection through Istanbul. And while, in my case, the flight from Lebanon to Turkey is painless, the run from Istanbul's Atatürk International Airport to Tashkent goes on and on. Sitting as they do, roughly 2,600 miles apart, Istanbul and Tashkent are much more distant than one might first imagine. After all, you have to cross the storied and endless plains of Central Asia, and as a Westerner, while this second part seemed like a jaunt between two exotic Eurasian capitals, side by side on the map in my mind, they're actually almost as far apart as Los Angeles and New York.

But really, the problem encountered comes from something else. Not long after takeoff from Istanbul in a Korean Air 747, there is an announcement from the cockpit. For reasons that defy understanding, a flight officer says, the bathrooms appear not to be completely functional. People must please refrain from using the lavatories unless absolutely necessary. And even then, if they do use the bathrooms, they shouldn't flush the toilets.

The aircraft is full of Korean tour groups, happy people headed back to Seoul and beyond from vacations in Turkey and on the Mediterranean. As a group, they all entered the aircraft a little before 9 a.m., largely relaxed and satisfied. Most of them are middle-aged and wearing buttons or laminated identification cards on lanyards. In a sort of caricature of Asian tourists, one of the groups aboard actually comes down the jetway following a tour leader holding up a small, red, pennant-style flag on a retractable metal pointer.

With the bathroom announcement, a sense of tension washes through the passenger area. Still, for the early segment of the coming six-hour flight, we all cope.

Then, perhaps halfway through the trip, one man I'd seen back in the terminal café in Istanbul emerges from a bathroom near my seat, followed by a tide of sewage.

The smell is horrible, and it hits the cabin all at once. The man seems shocked, confused. He's stepping lightly, trying to keep his shoes out of the mess. He hustles to the galley and informs a female flight attendant of the situation. She leaps into action, employing absorbent towels and a far-smaller-than-usual mop. To try to mask the smell, the flight attendants drizzle ground coffee out of foil envelopes into the flood. The man, still befuddled, stands there, watching.

As this happens I am thinking of how, in closed systems, things sometimes go horribly wrong. The smell is inescapable. In the seats around me, a few people have started retching into the small white paper bags always supplied by airlines in the seat-back pouches. We are all packed into an airplane. There is no escape. For the next four or so hours, 400-some of us are stuck here above the earth, inside a flying cesspool.

I try to breathe through my mouth, and return to watching a movie on a small video screen embedded in the back of the seat in front of me. Kate Hudson is playing a woman who is trying to raise her deceased sister's children. The people who wrote the movie clearly are trying to mine the full spectrum of human emotion.

In the aisle behind me, flight attendants are still trying to clean up the bathroom mess. The confused man is still standing nearby, clearly feeling helpless and hoping to be of assistance. Then another Korean man, who seems to be the head of a tour group—he wears a lightweight yellow vest over his button-down shirt; the vest has an oblong blue label sewn over his left breast—walks down the aisle and, just past my seat, stops in front of the man who set the bathroom flood into motion.

"You a *bad man!*" the yellow-vested man shouts, inexplicably, in English. Then, with his open right hand, he slaps the confused man hard across his face, the impact making a hollow pop.

*"Bad man!"* the yellow-vested man shouts again.

The Korean man who has just been struck stands and stares at the man in the yellow vest, now doubly stunned. Around him, no one knows quite what to do. The flight attendants step between the two men, so no further violence will occur. They're trying to calm a bad situation; to get life moving more smoothly again.

So here we all are: stuck on this airplane, together.

By roughly 10,000 years ago, *Homo sapiens* had populated much of Earth. And overwhelmingly, the modern humans not only discovered farming, a fair percentage of them were also now actively working to cultivate the earth. They were living everywhere: in what is today China and South America and Indonesia, across much of Africa, and all those portions of Europe that at the time were no longer covered in the ice sheet that had blanketed the region for thousands of years. The only *Homo sapiens* not farming in some capacity were the Arctic, African, and Australian hunter-gatherers, who, it's been speculated, hadn't lifted their populations to levels that degraded their sources of prey. Most everywhere else, humans had figured out how to thrive on Earth. And they were doing it.

But just as their ancestors had understood, lives spent hunting and gathering were difficult, modern humans were learning that a farming existence could be difficult, too. Judging by assessments of the archaeological findings, the average life expectancy at Çatal Hüyük was roughly 29 years for women and 34 years for men. In terms of life expectancy, things were no better for early Neolithic farmers than they had been for cave-dwelling Neanderthals 25,000 years earlier. In fact, the average life span of these new farmers was possibly even shorter. Perhaps this had to do with the vagaries of seasonal weather and crop production, or the rise of viruses and the need for improved sanitation in settlements of larger, more sedentary groups. But while day-to-day living likely got easier thanks to that Neolithic revolution called agriculture, the reward of a longer life had yet to be enjoyed.

Still, there remained another persistent issue dogging the notion of farming versus mobile hunting and gathering. Though year-round springwater, arable soil, and protective walls had let people live in what is now Tel es-Sultan and the city of Jericho for years, building each new layer of society on successively higher mounds of archaeological evidence, at Çatal Hüyük the archaeological record tells another story. On the plains of central Turkey, deforestation, subsequent erosion, and overgrazing appear to have rendered another of the world's oldest settled sites less and less viable. The trees providing the timbers that supported each of the honeycomb roofs, the actual streets of the town, were being used up. The earth had grown tired. Perhaps ample stores of water had become an issue, too.

At Çatal Hüyük and other early Neolithic agricultural sites, modern humans had created a garden and then watched it deteriorate across just a few dozen generations. By around 6,000 B.C., or 8,000 years ago, many of the first Neolithic agricultural sites had been abandoned. Beneath what archaeologists believe may be 13 successive layers of buildings, the inhabitants left behind sophisticated pottery and scraps of woven textiles. They also left behind kitchens, with human burial plots intentionally placed underneath the hearths of homes; perhaps to keep the dead warm during the cool of winter. They left behind terracotta figurines in the warehouses and cribs, likely images of deities meant to protect their stored crops.

And yet, the people themselves were gone.

Tashkent is Central Asia's largest city. And despite the endless-seeming plains around it still being home to hundreds of thousands of the nomadic peoples who've lived there for thousands of years, the city itself is modern and sophisticated. There's a network of metro rail lines, and the shops sell the same clothes and sunglasses one might find in Paris or the San Fernando Valley: a list of items extending to such things as soft ice cream and shiny new imported cars. Looking at the city, except for the Cyrillic writing on many of the road and commercial signs, Tashkent could be anywhere.

Deplaning the difficult flight to Tashkent from Istanbul (the Seoul-bound crew and passengers were met at the airport with a new Korean Air 747, just flown in and readied),

I get through customs, ready to prepare for the following day's overland trip via hired car to Samarkand. As I wait in the baggage area, I learn that the journey has not been without snags. The only other person getting off the flight here in Samarkand, another American, has lost his two suitcases, and, feeling lonely for American company, I try to help him negotiate the problem. "They're gray Samsonite suitcases," we keep telling the baggage-claim people at the airport. "Here are the baggage-claim tags."

The people who run baggage claim at Tashkent ask us to visit a small and airless office on the floor below, where the other American fills out forms. They tell us that, because the Korean Air flight was not a usually scheduled stop but was instead a chartered flight that was only landing at Tashkent to refuel on its way to Seoul, well, there's little they can do to get the lost bags back any time soon. Korean Airways doesn't have daily service to Tashkent.

The officials ask for the man's home address, back in New York, and suggest that eventually—thanks to computerized records and linked airline partnerships—the bags will be returned. "They are here, in the system," a woman in the baggage office at the airport keeps saying. "But I don't think the bags will return to Tashkent. No. I think not at all."

I take a car from the airport to my hotel. And find it's surprisingly easy to like Tashkent.

The summer weather is high-desert perfect: Beneath a sharp blue sky the day is warm but dry. My hotel, the Radisson, although it is downtown, has a huge rectangular swimming pool ringed by a chair-dotted concrete deck, with eight-foot squares of bright red or yellow cloth pillows on the outlying grass that are available for visitors to rest on. The Internet connection in the hotel works great. On my room's TV is *The Gathering Storm,* a BBC production about Winston Churchill's so-called wilderness years, that stars Albert Finney and Vanessa Redgrave. For dinner, just for fun, the concierge makes me a reservation at Jumanji, a local restaurant that's decorated with authentic African masks and art and furniture: It serves what's said to be among the city's best Chinese food.

After dinner, following a cone of soft ice cream that I worked on as I walked back to the hotel, I check my e-mail. My wife and kids have sent digital photos of themselves waterskiing at my in-laws' house in Arkansas. Everyone, including our chocolate and black Labrador retrievers, looks happy to be on the boat, enjoying their summer beneath the Arkansas sun.

Sometimes, the world offers nonstop amazement.

The next day at 10 a.m., a hired car is waiting outside the hotel. Time for the roughly 200-mile run to Samarkand.

It's a lovely morning, and the streets of Tashkent are full of people in clean and pressed clothes, heading to work. As we begin our journey beneath yet another high-desert blue sky, my driver, who has introduced himself as Golbor, steers us slowly out of town to the southwest. "So . . . why do you visit?" he asks. His English is quite good.

I explain. We roll alongside a metro line and cross and re-cross a river. We pass the university, which Golbor points out.

*The Registan in Samarkand, ancient center of learning and trade*

"Yes . . . yes," he says. "If you want to meet Tajiks, then you must go more to the south. Here in the north, we are Uzbeks. Even the language is different there. They speak Tajik, which is a kind of Farsi, or Persian. We speak Uzbek. . . ." Then he goes quiet.

And almost all at once, the city is behind us and we are into the plains of Central Asia.

As advertised across the tales of history, the steppes of Central Asia are huge. They spread in every direction, gray and—truth be told—more than a little monotonous. For a time, canals and dikes run near the road, their pipes and headgates used to direct water to the surrounding fields, which appear to be growing cotton or vegetables.

In the distance near the horizon, like some sort of high school art project on perspective drawings, I occasionally spot a huge, distant, blocky factory complex near the flat line where land meets sky, often with a plume of smoke coming from a smokestack.

According to my guidebook, this part of Uzbekistan is called the Hungry Steppe. And it appears famished.

On the two-lane highway, our Daiwoo taxi hurtles along, passing full buses and pickup trucks whose rear beds are filled with workers. Every 15 or 20 miles, we enter another ragged-looking little village, usually made up of small, wooden dacha-style houses, remnants of the Soviet era, their window shutters painted various bright colors. The houses are homey and kind of beautiful, though awfully poor-looking.

In about an hour, we pass a turnoff for the city of Guliston, where a huge building hulks on the horizon. Black smoke pours from the top of the stacks and heavy power lines run to it.

"What's that?" I ask.

"An electricity plant," Golbor says. "From the Soviet times."

The car goes silent again. Golbor does not even play the radio. Of course, given that this place is so isolated and barren, maybe there aren't radio signals to be gotten out here?

Three-quarters of an hour farther along, a line of flinty and eroded mountains appear to the south, and about the same time the main road makes a slow but discernible turn to the west, and a small, secondary road points due south, toward the ridges. "What are those mountains?" I ask.

"That is the Turkistan Range," Golbor says. "Those mountains, they are in Turkistan. The road, if we were to have gone straight, would have led us into Turkistan in just a few kilometers . . . in very little time. This part of our country is a narrow neck for Uzbekistan."

Golbor hooks his right thumb directly behind him, pointing out the back window. "Just behind us, closer even than Guliston, is the border between Uzbekistan and Kazakhstan," he says. "A road goes that way, too. I don't know if you saw it. But if you have the right paperwork, you can travel in three countries in about an hour."

Then there's more driving in silence.

Our taxi rips along. Now the road is virtually empty, as we've gotten in front of most of the slower-moving buses and trucks carrying people out of the city that departed about the same time we did. Up ahead, rectangular carpets of many different sizes have been laid across the road, and Golbor slows the car slightly but cruises right over them. This happens more than once.

"What's up with that? Carpets in the road?" I ask. "Did they fall off the back of a truck or something?"

"Oh no," Golbor says. "Those carpets were put there on purpose, by the herding people. These are new carpets, which herds-people have woven themselves from the wool on their sheep and colored with natural dyes and sometimes also spun with a little silk, if they can get silk from the cities or surrounding villages. As you probably know, new carpets are not as valuable as older ones. So these people sometimes put carpets onto the road and let the cars and trucks run over them, to make them look older and more worn. It brings more money."

This is true, as I later confirm in my copy of the bible of tribal carpets, *Oriental Carpets: From the Tents, Cottages, and Workshops of Asia,* by Jon Thompson. In this book Thompson explains that, for millennia now, the tribal herdsmen in Central Asia have taken two of the materials most critical to their lives—wool and leather—and turned them into an art form the world now craves to the tune of billions of dollars in trade a year.

Interestingly, only the pastoral tribes in the west of Central Asia developed the "knotted pile" technique that today makes up much of the carpetmaking world. Many believe these so-called Oriental carpets were woven in patterns that might imitate the texture and insulating qualities of animal pelts, which were probably more valuable. Amazingly, the oldest known remnant of carpet was found in the frozen tomb of a nomadic chief named Pazyryk in southern Siberia, and it is thought to be from at least 7,000 years ago. Then—for almost 17 centuries—the subject goes dark.

Except for a few other carpet scraps found in ruins in eastern Turkistan or unearthed from ancient middens outside Cairo, all news of knotted carpets disappears until the 13th century, when they made it onto the earliest inventories of mosques in Central Asia and the Bosporus region.

So while carpetmaking didn't cease across all this time, it wasn't until the 13th century—during the Seljuk empire—that knotted carpets emerged from their anonymity out on the Hungry Steppe and thrived on the world's stage, becoming beloved by European royalty for their "authenticity of character." Even King

Henry VIII hoarded them. More than once, he had a portrait of himself painted amid his collection of rugs.

Just ahead, the highway slips through a narrow rock gorge, barely visible from a half-mile away, and begins down into a valley, the dry gray soil replaced by yellow-colored stone and small green trees that look like elms. A small, dry creek even runs at the gorge's base.

"This is where, it is said, Alexander the Great came through on his way to Samarkand," Golbor says. "This is supposed to be part of the Silk Road. Some people call it the Gateway. Not too far away, maybe 40 minutes more, there is also a place that it is said the Old Testament prophet named Daniel is buried. We are almost to Samarkand now. In another hour or so."

As Golbor promised, before long the road traffic has grown more congested, and the pointy steel tips of radio towers poke into the sky above the horizon. Golbor motions ahead, toward the brown outline of an eroded hilltop encircled with trees south of the road. "That hill there?" he says. "That is the site of the Ulugh Beg Observatory: from the 15th century. We are at the edge of the city. You have arrived."

Thirty-five hundred years ago, Samarkand was only beginning to rise up as a trading center. Set in a relatively verdant valley between mountain ranges—geological features that funnel the otherwise broad, Central Asian plains toward it—the city's location remains a natural focal point for anything moving across the region.

"You know, no one can really say why some cultures, or some places, rose up in the ancient world and others didn't," says archaeologist Alison Brooks. "It's probably a very different set of circumstances in every case. But probably the specifics of the landscape and a flair for adaptation or invention by the indigenous people were both large factors."

Which is likely true for Samarkand, though it would take several hundred years before the city's full power as a center of human affairs reached its peak. At the same time, the kinds of felicitous events and circumstances that were beginning to raise Samarkand from the earth in the form of mud huts and a few traders laying wares on the ground; elsewhere, unique characteristics were also beginning to raise societies up as well.

A thousand miles to the southeast, for example, in the land between the Tigris and Euphrates Rivers—a place that would come to be called Mesopotamia—a farming paradise—had started to emerge at the same time as Samarkand.

On rich alluvial plains fed with water-borne minerals of the two rivers, an organized farming-based society took shape. Set in a network of marshy lands rich with wild animals and waterfowl, a collective of human effort and forward-directed thought elevated the Neolithic revolution's idea of farming through networks of hand-dug irrigation canals.

"This natural paradise required intensive labour and organized co-operation of large bodies of men," the famous archaeologist Gordon V. Childe wrote of Mesopotamia in his book *New Light on the Most Ancient East,* "arable land had literally to be created." This was accomplished "by a 'separation' of land from water; the swamps must be drained; the floods controlled; the life-giving waters led to the rainless desert by artificial canals."

With ample water and rich soil nearby, not to mention simple trade for tools carried by travelers passing through, Mesopotamia's farmers were building a foundation for their world: a society that would thrive for several millennia. Though they were still without the strong durable metals from which to make tools, and irrigated farming remained demanding work, the thoughts and techniques behind agricultural technology were growing more sophisticated, as was the ability to grow ample crops in the rich soils of what is today southern Iraq. Yet because it required less time and energy to feed themselves, societies in Mesopotamia began to diversify into different groups or skill-sets: There were cultivators, irrigation workers, makers of mud bricks for houses, and even some early administrators. Many observers who came later—Gordon Childe and Alison Brooks among them—believe controlling the flow of water to tended crops led to the rise of the first early civilizations.

Yet even as they thrived and watered and harvested—and may have even created the first pottery wheel—the farmers of Mesopotamia weren't alone. Slowly, at about the same time, along the Nile River in northeast Africa, people had begun to irrigate and cultivate. They were also watering in the lands called Elam, in what is today central Iran, and along the enormous Indus River in modern-day India, and in the Zana Valley of Peru.

Like the rise of subsistence farming 3,000 years earlier, the establishment of larger-scale farming through irrigation was equally a global phenomenon that seemed to emerge in several different locations at roughly the same time. And though travel, trade, and cultural and technological interchange were possible in the swath of land between the Nile and Indus Rivers, the rise of irrigated cultivation didn't stop there. Most notably, in what is today north-central China, where deep layers of rich aeolian loess and alluvial floodplain soils had accrued across millennia, the Yang-shao and Long-shan farmers were also learning to irrigate and cultivate, despite having no provable interchange with the people of the Mesopotamian world. They, too, built pressed-mud buildings, and they had developed clay pottery, some of which shows remarkable sophistication, featuring pouring spouts and handles.

Across the world, from what began as scatterings of mud-brick huts in proximity to irrigation canals and farmland, cultures began to evolve and diversify. Just as happened at Tel es-Sultan and Çatal Hüyük, these early settlements improved and rebuilt upon themselves, birthing archaeological mounds beneath their footpaths and foundations. They became *places*, and subsequently were given names like Babylon, Susa, Ur, and Uruk in Mesopotamia; Memphis and Buto on the Nile; Mohenjo-Daro along the Indus; and Taosi and An-yang in China.

Irrigated farming as a cultural stabilizer became an idea that spread as if carried on the wind. Yet even more than the farming practices and the development of rudimentary pottery, the farming societies found other common features at the same time. Thanks to the collective work (and self-interest) that irrigated farming promoted, it is thought—and can be demonstrated to a degree through the archaeological record—all of these societies developed primitive political systems. These afforded ways to address questions like: Why should one crop, or one area, get more irrigation allocated to it than another? And how can we make this equitable?

To accommodate these earliest governmental institutions and their public requirements, common areas were set aside, where people could meet and air their ideas; then came larger public squares and buildings, social centers and meeting places.

Years passed. Then centuries. New technologies emerged slowly as modern humans gained experience and imitated one another. Still, farming grew more sophisticated. In China, tiered fields were being developed on mountainsides, to use gravity to distribute water better. Also, as more towns rose as well-fed populations grew, other good farming locations were discovered in the outlying districts.

Progress was not all gentle and pretty, however. Because of the increasing value of the crops and animals being grown, and of the wealth that different villages and cities and societies accrued, according to Sethard Fisher, Ph.D., of the University of California at Santa Barbara, the first security forces came into being, in the forms of local police and standing armies. These were developed to keep internal squabbles from getting out of hand and for protection against external robbers and raiders. According to Fisher, crime and economic development were now evolving together. And for those directing the armies, the kings and highly placed social officials, they soon must have realized they could use these forces for their own protection, as well.

Human society was growing up. And it was growing up fast.

Then, a little more than 5,000 years ago, there came something so new and important it matched in its magnitude the technologies of farming, division of labor, and pottery. It would become as big as the impact of political planning or the introduction of organized crime and dedicated security forces. As a way to demarcate whose land was whose, to provide accounting for posterity, and to give direction to those who needed it, the first early symbolic pictographs, the first writing, appeared.

Many scholars believe the idea of tally marks probably began in Africa. There, several "tally bones" have been discovered, and have been dated as far back as 9,000 years. The most famous, called the Ishango bone, the fibula of a baboon etched with tallies that add up, over and over, to prime numbers, shows that people had long been counting with some sophistication. Still, while tallies may have started in Africa, the idea of lasting written symbols seems to have first emerged in Mesopotamia. And to accomplish it, the Mesopotamians employed clay tablets that could be etched with agreements and then dried to hardness in the sun. It's also believed by some scholars that the Mesopotamians and others had been cutting symbols in the earth near owned fields and at crossroads for some time before they realized a fixed surface with writing might be a good idea.

Anyway, when prepared, each new clay tablet could be etched with a sharp stone wedge or a reed, a tool that would come to be called a *cuneus* in Latin. From that, an understandable vocabulary of symbols began to take place, lines that would eventually evolve into a repertoire of denoting marks that became numbers and, later, letters strung together to encompass words . . . and thoughts.

At first, though, these tablets appeared as a form of agricultural recordkeeping. Because the farmers of Mesopotamia kept their harvested crops in communal storage areas, the tablets memorialized recorded transactions, and were often kept in clay jars: the first ledgers. Not surprisingly, these tablets over time became the first money. Suddenly, "pay the bearer of this object" credits came into existence, though, in terms of finance, humans still had a substantial way to go. As the Harvard historian and author Niall Ferguson writes in his book *The Ascent of Money:* "It would not be quite correct to say that credit was invented in ancient Mesopotamia. Most Babylonian loans were simple advances from royal or religious storehouses. Credit was not being created in the modern sense. . . . Nevertheless, this was an important beginning. Without the foundation of borrowing and lending, the economic history of our world would scarcely have got off the ground."

Suddenly people were storing resources and trading against the idea of existing assets, and lending against them to other individuals or institutions. And thanks to routes of trade that brought interaction with other cultures across Asia and beyond, a notion of symbols and written language, along with the idea of extended credit and money, began to spread. *Homo sapiens* ("wise man," as he would later dub himself) was beginning to live up to his name.

By the time of cuneiform's invention, farming and the new sedentary human cultures that grew food and kept livestock for meat and milk were stabilizing life for a large and still growing portion of humankind. With the addition of trade in credit or clay tokens that would eventually become money, that stability would be not only reinforced but magnified. Money and written language would come to reflect human nature—and could document how cultures continued to move forward. And with cuneiform propelling it all onward, the velocity of cultural progress now being recorded by the world's dominant species was about to rocket ahead.

My hotel in Samarkand is called the President. Located near the city's center, it was built as an Uzbek–German joint venture and is said to be the nicest place for visitors to stay in the city.

As Golbor drives us into town, what is immediately noticeable is that—like rings on a tree—Samarkand is a place where the new encircles the increasingly older. At the outer edges of the city are contemporary, villa-like houses from the Age of Reality TV, complete with multigabled roofs and walled swimming pools, apparently the fruit of a recent economic boom. The next ring inside is filled with Soviet-style high-rise apartment blocks of poured concrete, each maybe 12 or 15 stories tall. Despite tree-filled parks and green common areas between the skyscrapers, with their rows of dark windows and clearly enforced sameness, they throw a gray-faced visage on the world. The next circle is populated with lower, two- or three-story Soviet-style buildings, also of poured concrete, with grassy lawns and Bauhaus-style windows that present a more welcoming face.

Then we're downtown, in a city of ancient warehouses of mud-brick and squared-off mosques with tiled exteriors and soaring minarets. Every half mile or so, the buildings relent, opening onto large pubic squares: places that could hold thousands. There is a simple trolley system on the main road. The streets are wide, with lanes of opposing traffic often split by trolley tracks and park-like strips of grass or spreading shade trees beneath a sunny sky. This is the ancient city center.

Just ahead and off to the right, a tricolored monolith stands in front of us: yellow plaster, blue-coated glass, and brown-marble make up what is, by far the most imposing building in the neighborhood. It's my hotel. When Golbor pulls into its circular driveway, he puts voice to the obvious: "Now . . . we are here," he says.

Before the car has stopped at the hotel's entrance, a cadre of greeters is coming out to help us. Some of these men hold open the hotel's glass-paneled doors, and others want to assist with my baggage. They are nothing but eager, whisking my duffel bags from the Daiwoo's trunk and into the hotel on a tide of energetic movements, quick smiles, and small, nodding bows of the head.

I pay Golbor for his work and, that fast, he is headed back to Tashkent. The hotel's interior has a soaring, atrium-style lobby, with guest hallways ringing each floor all the way to the glass-paned ceiling. It could be any hotel in downtown Atlanta or Dallas. But there's a second thing truly notable about this seven- or eight-story lobby: It's devoid of people. It's empty. There are the doormen and the desk staff behind a marble and wood counter, but the hotel is bereft of visitors.

"Welcome . . . welcome," the manager says. "You are Mr. Webster. Here as a guest?"

"Yes."

"Good. We have been waiting for you."

In minutes, my Visa card has been processed, my passport has been copied, and I am in a corner room on a middle floor, which in its institutional sameness could also be a hotel room anyplace in the world. There is an ample desk for work, complete with electrical outlets for laptop computers and cell phone chargers. There is a big color television. There is a small refrigerator, and a well-appointed bathroom. On the far side of the bed are big windows overlooking the hotel's park-like grounds and the city beyond. Though it's been only a few days since I left Tanzania, I'm now officially far, far away from hunting and gathering.

Once again, my new surroundings and the people living in them are nothing short of entrancing. Within minutes, after passing for a second time through the deserted hotel lobby, where everyone on the staff wishes me a good day, I'm on the street, ready for a walking tour of Samarkand. And already the city is drawing me in.

Just heading up the sidewalk toward the historic district, it's impossible not to notice the population's diversity. The people here are everything all at once. Heavyset or fine-boned, with Asian features, Euro-Russian features, and Indian features, each seems to tell a personal story of the cultures that have passed through or conquered this place.

There are also little kiosks everywhere, where Cyrillic-lettered newspapers and magazines, soft drinks, batteries, and butane lighters are offered for sale. And each kiosk seems to be doing brisk business.

Following the map in my guidebook, I take a right on a street called Akunbayev, walking as I read about the city's history. With a location providing both ample water and a geological bottleneck for travelers moving across the plains of Central Asia, Samarkand quickly became a natural capital of trading. And much like the excavations at Tel es-Sultan and Çatal Hüyük, archaeological digs here have uncovered layers and layers beneath the city's streets, buildings, and sidewalks, with coins from both ancient China and Rome having been unearthed amid the rubble of ancient walls on which vestiges of frescoes are still visible.

Just ahead is the Gure Amir Mausoleum. A sprawling building with carved minarets flanking its arched entrance, preceded by a tall arched entrance gate faced with blue tiles and topped by a fluted, shiny, pale-blue glazed dome that glimmers in the sun: This is the burial place of Timur (or Tamerlane), Samarkand's leader and the reigning tribal warlord of the 14th century. As I approach the building, a skinny, black-haired, and slightly Asian-looking man in his 20s steps toward me. He's wearing a white T-shirt, tight black jeans, and low, red canvas sneakers. Despite the afternoon's glare, his sunglasses are perched atop his head.

"Hello visitor," he says in English, his voice full of bounce. "I am Dilshod. I will be your guide." He flashes a plastic-laminated card at me, apparent evidence that he is official with regard to Samarkand's tourism industry.

"No, no . . . really," I say, waving him off. "It's okay."

"Oh, you need a guide." Dilshod is now walking with me as I try to step briskly past him. "I will tell you things you cannot know otherwise."

"How much will it cost?"

"We'll decide at the end of the visit," Dilshod says. "Maybe my pay will be nothing. Maybe it will be a nice amount. We'll decide later."

"I don't like the sound of that."

"You will not be disappointed."

As we walk through the gate, across a small garden area, and to the front of the mausoleum. Dilshod keeps talking. "I am a lifelong local in Samarkand," he's saying. "I know things others do not."

By this time, we're at the mausoleum door, which feels like a small niche inside the entrance's 20-foot-tall recessed archway, covered in the blue tiles touching edge to edge that I saw from the street. I pay roughly $2 to enter, thinking this cost will keep Dilshod outside, but he walks into the mausoleum right behind me. "These guys at the door, they're my friends," he says.

Then Dilshod launches into his story. Though Timur didn't want to be buried in Samarkand, this is not the way things worked out. A Central Asian herdsman and warrior to his core, Timur, a huge and imposing man who Dilshod asserts is "still the ideal" of all Tajiks, desired to be buried out on the steppe. In fact, at the time of his death he had already ordered a small and virtually anonymous crypt to be built for him in Shahrisabz, the place of his birth, about 50 miles to the south. This mausoleum in Samarkand is a far more imposing building than Timur wanted for his own burial, Dilshod says. It was built by Timur in 1404 to honor his family and grandsons, since, by then, Timur had warred and sacked enough of the outlands to become very, very rich. With this wealth, he decided to turn Samarkand into his people's capital city. It survived for thousands of years, largely thanks to its fortuitous location. "Still," Dilshod says, "Timur wanted to be buried in the humble location of his childhood."

Then tragedy struck: Toward the end of 1405, as Timur was in what is now Kazakhstan preparing a warring and pillaging expedition into the lands of the western Chinese, he contracted pneumonia and died. Because the mountain passes back to Shahrisabz were already blocked with early winter snow, Timur had to be interred in Samarkand. Buried near him are several descendants and a favored teacher.

"It's funny sometimes, how the life goes," Dilshod says.

Truth be told, taken as a whole or in little snapshot bits, the mausoleum is a magnificent thing. Its exterior is tan and blue, the brightly pigmented and shiny glazed tiles affixed to its outer walls flashing in the sun with each step you take. The interior gleams with more blue fired tiles decorated with gilded accents. Jade inlays in many of the tiles give a slightly green cast to the air around them. Despite the imposing look of the mausoleum from the outside, once inside you see that it's not a big room. And though the markers denoting the deceased look like tombs, the bodies of those honored here are actually interred one floor below. The marker for Timur, Dilshod says, is made of extremely dark green jade.

"That kind of jade very rare and unusual," Dilshod says. "But that is not my favorite fact of this place."

In June 1941, back when Uzbekistan was a part of the Soviet Union, a famous Soviet anthropologist named Mikhail Gerasimov wanted to examine the burial crypts of Timur and his family. After he got permission, he opened the graves. Inside, he found that one member of Timur's family, Ulugh Beg, a grandson of Timur and a longtime ruler of Samarkand, had been beheaded.

Gerasimov also found that, fitting with oral tradition, the skeleton of Timur showed he was tall and quite physically imposing, but that he had been seriously injured along his right arm and leg, leaving him slightly hobbled. "In fact," Dilshod says, "Timur's other and more Westernized name, Tamerlane, is actually a version of 'Timur the Lame.' So that also fits with history. But what is most amazing came next. In examining Timur's grave, Gerasimov found a carved tablet with this warning: 'Whoever disturbs this tomb will bring the demons of war upon his land.' "

We are alone inside this beautiful, almost intimate building, with its elaborate blue and gold inlaid tile walls and jade latticework. I look at the almost-black jade box that marks the passing of a great man 600 years ago. As I glance around, Dilshod stays silent for a theatrically long moment. He is smiling.

"Okay, so the day after Mikhail Gerasimov opened the tomb of Timur?" Dilshod finally says, "That is the day Adolph Hitler invades Russia. He invaded with three million soldiers. Across history, the German invasion of Russia in World War II is the largest military assault of all. Before it was over, with the Germans finally driven back and out of Russia, between 20 and 27 million Russian soldiers and civilians were dead from the invaders. The cities of western Russia had been reduced to smoldering ruins. Across the countryside, women and children could be found hanging dead from the eaves of their houses and from electrical poles, ropes around their necks, left there by the Germans. A large percentage of all the men in Russia were dead from the fighting. It was a landscape that must have looked like hell."

Dilshod shrugs. He smiles and tosses his hands in the air. "But, you know," he says. "All of this is probably a coincidence, right?"

Leaving the mausoleum, I head for the city's top tourist attraction, and one of the great sights of Central Asia, with Dilshod on my five o'clock as I go. Called the Registan, my next stop is said to be an ensemble of three enormous, majestic, slightly age-worn buildings perhaps half a mile distant. It's a short trip: I take a right, then another onto a broad boulevard called Registanskaya.

Dilshod offers to come along.

"No, no, it's fine."

"No, really. You are interesting."

"Yeah, right."

There's no point in trying to lose this guy. Dilshod has now latched onto me, and he probably has an all-access pass to any tourism site I'd want to visit this afternoon. In fact, in the future, any time I think of my time in Samarkand, a mental image of Dilshod—in his slightly ill-fitting clothes and mop of black hair above an ever present smile—will flash across my mind.

As we head for the Registan, I explain what I'm doing in town.

"Oh . . . so you are part *Tajik*?" Dilshod asks as I finish explaining. "I'm Tajik, too. Perhaps we are long-ago related?"

Dilshod goes on to recount the history of the Tajik people, with—I will later learn through my own research—a fair bit of accuracy. The Tajiks are Farsi-speaking tribes that are descended from people who emerged to the northeast from Persia (modern Iran) about 40,000 years ago; their more modern tribal designation literally means "non-Turk." And they have been living on the plains of Central Asia ever since. Consequently, their lineage is enormous, creating the foundations for what population scientists often call the Eurasian Clan: a group so large and far-flung that most people living in the Northern Hemisphere today have Eurasian Clan markers in their DNA. And mine, *M9, M45,* and *M207* leave me firmly attached to the Tajiks, no matter what my more recent family documents say.

These people were successful as hunters and accomplished as toolmakers, and the archaeological record even shows that, from their beginnings about 35,000 years ago, Eurasian Clan members began to develop more sophisticated tools, such as sewing needles made of bone, which they used to stitch together warm leather clothes and leather shelters against the region's sometimes brutal weather. Ultimately their wide distribution across the Northern Hemisphere is believed by Spencer Wells and others to have been precipitated by an ice age that peaked about 20,000 years ago and lasted an additional 10,000 years. The approaching glaciers across the Central Asian plains left prey and other resources hard to find, forcing the population to band into small, mobile groups in order to survive. These groups spread out, scraping out a living wherever the environment would allow. "We Tajiks have been herdsmen and nomads and light farmers ever since," Dilshod says.

We round a long, slow turn in the road, and there, ahead of us behind some trees on the other side of the street, stands the Registan. Its huge public square is wrapped on three sides with gigantic, box-like, domed buildings flanked by minarets.

"Wow," I say.

"Yes," Dilshod says. "The Registan. It is big. Big. Beautiful."

Massive in its scale and stunning in its decoration, the Registan not only has towering minarets, its screen-like *juli* walls are carved with a thousand little breeze-providing slits, color-splashed spandrels, and depressed entrance archways, all covered by geometric designs in blue and gold and white majolica tile. As a sight, it's nothing short of jaw-dropping. Both ancient and eerily contemporary-feeling, Islamic and classicist at the same time, the Registan is gigantic yet exceptionally well proportioned. In fact, because of these perfect proportions, the whole arrangement seems closer to me than it actually is. For several minutes, we keep walking, the buildings growing larger and larger as we approach them. Then, it takes us several more minutes to get into the center of the Registan's public square. And yet, the whole time, I never felt overwhelmed by the size of the structures and the vastness of the space around them.

"Today this is still pretty much seen as the center of Samarkand city," Dilshod says. "And it has been this way for a long, long time. Far longer than the 600 years since these buildings were constructed."

As we walk, Dilshod keeps talking. He tells me that all three of the buildings were Islamic schools, madrassas, and that the oldest of the three buildings, the Ulugh Beg Madrassa, off to our left, was completed in 1420. Ulugh Beg, Timur's grandson and a beloved leader of Samarkand, lectured on mathematics in one of its original four chambers, which sit at the interior corners of the building.

"Records show they taught other things here, also," Dilshod says. "Those subjects included philosophy, theology, astronomy, . . . history. The other two madrassas were erected in the 1600s. But this place, as I say, is far older. When the Russians came in and decided to restore the Registan in the 1980s, they dug down to the buildings' foundations and they found debris *three meters* deep here in the central square. Did you notice, as we walked over here, how the ground slopes down? Suddenly, with the Soviet restoration, these buildings grew three meters *taller*. Inside that debris pile, the Russians found everything from the ancient world; a record of what had come before. They found pottery and sand and bones and small bits of tools."

Dilshod keeps talking, but I'm only half listening now, still a little stunned by the scale of this place. He's telling me that, though Timur died long before the Ulugh Beg Madrassa was finished, he used the square, which had long been the city's center, as a place of executions. After these public beheadings, the executed people's heads were left in the square, impaled on spikes. It's also alleged that, to soak up the blood at the end of a day of executions, sand was spread around the square. Dilshod also tells me that, whether or not those rumors are true, the archaeological record shows that the square fronting the buildings was probably a main market for the region for thousands of years.

"Can you imagine it?" he says. "People selling food—lamb and bread and yogurt—next to human heads on spikes? What a *vision* of the world. Of course, this may all be legend. But there are things that the archaeologists can show us, evidence that can prove the city was a trading center on the Silk Road even in its earliest days. They have found bits of silk and bronze, and stones like lapis assumed to have been transported from as far away as Afghanistan, because they are not found in the earth here. And these seem to have been here for thousands of years before these buildings arose. Though there are people who disagree about when the Silk Road trade started, I think you can at least say that this place was, for a long, long time, one of the centers of the known world."

---

Perhaps as much as any other human invention, trade built and advanced the modern world like nothing that had ever come before. Excavations in China, in lands south of the Yellow River, thousands of miles to the east of Samarkand, for example, have regularly unearthed cowrie shells estimated to date back to over 3,000 years ago.

These are relatives of the shells used for the eyes in the skulls back at Tel es-Sultan. But in the Chinese excavations, they are sometimes in groups and found in containers, sometimes in public areas where early commerce might have been conducted. And the particular cowries in these collections appear to have come from the Maldive Islands, even more thousands of miles southeast of Samarkand . . . out in the Indian Ocean.

How did cowries from the Indian Ocean get to China? That's simple: They arrived through the rise of travel and trade.

Starting about 6,000 years ago, and continuing up to this moment, as humans became sedentary farmers living in stratified societies where labor was divided, some *Homo sapiens* became specialized traders and travelers, as well. Modern humans had learned quickly that, in a farming culture, specialization was essential. After all, if a day was spent farming, it couldn't be directed toward going out and looking for volcanic glass from which the stone knives were made. Similarly, if a man was an army commander, he didn't have time to make his own swords and grow his own food. In a fairly short stretch of time—just a few thousand years—geography and skills at production took central places in the world of multilateral commerce, with comparative advantages of different populations and cultures benefiting people in disparate areas along routes of trade. Wheat seeds moved from the Levant to the east, as the earliest silks and some forms of early pottery moved west. Over time, other things began to move, too. Eastbound traders carried gold, precious metals and stones, textiles, ivory, and coral toward China. Westbound traders often transported furs, ceramics, cinnamon bark, and rhubarb toward Baghdad.

Though it was called the Silk Road, it turns out silk was a relatively small portion of the items traded along its length. And it also seems that few transport caravans actually transited the entire route. Instead, they moved only city to city, trading and depositing their wares in the route's network then turning for home.

So though humans had become more settled in their day-to-day efforts in different locations around the world, that didn't mean human *ideas* had. With these travelers and traders, beliefs and practices and theories began to move around, too. And while other human societies were beginning to thrive in the shadows of retreating glaciers in Europe, as well as in the Americas, without the archaeological evidence showing any huge spikes in population, in the Middle East and Central Asia, *Homo sapiens* were meeting success like nowhere else on Earth.

In following another set of genetic markers, *M175* and *M122,* Spencer Wells and others can see that, arising in northern China (or perhaps even the Korean Peninsula), these markers and their carriers became widespread in China by 7,000 years ago. They spread and thrived in populations down the Yangtze River Valley. By 5,500 years ago, *M175* was on the island of Taiwan, and would soon be migrating to Borneo, Sumatra, and the islands of what is today the Indonesian Archipelago. And using the population record carried in DNA, plus seed evidence found in clay containers and structures of the age now uncovered by archaeology, it is thought the spread of this specific genetic marker is due in large part to one thing: the cultivation of rice.

"Looking at the pattern of Y variation in modern Chinese populations, it is now clear that the first agriculturalists in China were descendants of *M175*," writes Wells in *The Journey of Man*. "In fact, over half of the entire male population of China have Y chromosomes defined by a marker that shows evidence of a massive expansion in the past 10,000 years."

Because *M175* and a related marker, *M122*, are now found widespread across eastern Asia, but occur only rarely in Europe or the Middle East, Wells and his colleagues have come to believe "the development of rice agriculture in East Asia created a Wave of Advance." Man was farming with increasing success in more far-flung places; he was traveling and communicating and trading. His influence was spreading across the world. And it would only continue to grow. While Spencer Wells, Luca Cavalli-Sforza, and whole armies of paleoethnographers believe that at the time of the Neolithic revolution the world population numbered something in the low millions, by circa 1750 B.C., just 7,000 years later, that population would be 500 million. And between 1750 B.C. and the start of the 21st century A.D., thanks to agriculture and the rise of a thousand other technologies that render life both easier and lengthier, the population of the Earth has reached 6.5 *billion*.

However, while agricultural revolution sounds like nothing but a winning idea, all of this progress also carried with it a boatload of new problems. In tying themselves to farming, modern humans had to deal with land that would sometimes develop unexpected impediments through exhaustion and salting by irrigation. They had entered a world with technology traps at every turn.

The examples litter history. Easter Island, in the central Pacific Ocean, once full of trees and food, is now barren and full of tall volcanic-rock statues with humanlike heads staring out into the ocean. This resulted from a practice of raising more and more of these elaborate stone images to the memory of tribal and clan leaders on the island's *ahu,* or altars. Transporting each statue required more trees to function as skids pulled by ropes made from the woven bark of trees. Eventually, the people of Easter Island could carve more statues, but they could no longer move them, because the raw material for transporting these huge statues, the trees, was exhausted, gone.

Even more mysterious is what happened to the natives of Machu Picchu in Peru; their buildings and farming plots are still there, but their culture has vanished. Despite nearly a century of archaeological study, nobody has been able to fully describe what happened.

But, for our purposes at the moment, no example may be better than the ancient Mesopotamians. Having built themselves a paradise in the Fertile Crescent between the Tigris and Euphrates, complete with societal divisions and a government and the emergence of a written language, they encountered a problem they hadn't expected and never saw coming. There appeared to be lots of sediment in the rivers' waters, including just the tiniest bit of salt, carried downstream from the stones of mountains and hillsides upstream. These were the waters the Mesopotamians in places like Uruk and Ur were using to keep their fields watered and productive, the salt remaining in the soil as, year after year, the water evaporated.

Problems appear to have started with the wheat crops, which records show were replaced with barley, which had a higher tolerance for compromised soils. By 4,500 years ago, wheat had fallen from a large portion of Mesopotamia's harvests to a far smaller percentage of the annual yield. By 4,100 years ago, wheat had stopped growing altogether. And by 4,000 years ago, scribes were reporting that the earth had "turned white" from the deposited salt, with few crops growing well.

In Mesopotamia, modern humans had tied themselves and their growing society to one idea: irrigated farming. And because of the tiniest bit of salt—a thing inconsequential by itself but having incremental importance in the soil over time—Mesopotamia's irrigation technology was the source of that society's failure.

At the Registan over the next hour or so, Dilshod and I explore. We start at the Ulugh Beg Madrassa, passing between the two sturdy-looking minarets that flank its facade. Both are tiled in elaborate designs of blue and tan and red. As we slip inside the building through the enormous recessed arch of the front door, it is clear that Dilshod knows the security doorman.

The interior of the Ulugh Beg Madrassa is even more impressive than its exterior. And it isn't until I stand in the center of its main hall, a huge rectangle with soaring height, that I finally see the point of a mosque's domed ceiling. There, above me in tens of thousands of different bits of glazed blue and gilded tiles, concentric circles draw my eye up and ever inward toward the dome's interior peak. As my gaze follows those circles toward the ultimate, tiniest, interior center, perhaps 150 feet above, I begin to feel as if I'm being drawn up toward the peak's center, as well.

The effect is transporting, as if I've been lifted off the ground and am moving through time and space directly toward heaven. There's brilliance in its design. Decoration and the use of interior space has altered my perception in a way that I've never felt before. It's a completely different experience from viewing the soaring Gothic arches of St. Paul's Cathedral or Westminster Abbey in London, or the muscularity of the Eiffel Tower in Paris. This place has some sort of pure, transportive power. In the world, the interior of the Ulugh Beg Madrassa stands alone.

Beyond the central prayer area, the interior classrooms of the madrassa now house merchants selling souvenirs: silks and glazed platters and wall hangings and small models of the Registan.

"Sir, sir," the merchants all ask. "You would like a small carpet? A glazed platter? Some sandals? A few postcards to send home?"

Dilshod keeps the sales army away, speaking to each merchant in clipped, slightly harsh-sounding Tajik, his hands moving in knifelike slashes in front of him.

After the Ulugh Beg Madrassa, we move across the square to the left, or north, to the Tilya-Kori Madrassa. Completed in 1660, with its gold and blue exterior and pleasant interior courtyard gardens, it's clear it was once a place of peace and learning. Though the merchants are a little aggressive here, too, Dilshod keeps them at bay. Then, as we head back outside into the central square, the security guy at the door nods to Dilshod. He smiles. "Sir," the security man says, "would you like to see the Registan from the top of this minaret?"

He points to one of the needlelike towers flanking this middle madrassa. Its tiles are a little broken and peeling, and it seems to tilt a little, listing inward, toward the central square. "The view from up there is beautiful," the security guy says. Dilshod flashes a smile. "He says the cost is 2,000 Uzbek sum (about $2)."

The security man is not wrong, though the architects who designed the stairway to the minaret's top never planned for tall, wide-shouldered Americans. Still, seen from the top of the minaret, Samarkand spreads in every direction: a place of green trees and rolling hills and ancient tiled buildings and large public squares. Beneath a blanket of July sunshine, the view is lovely. It's obvious why people first chose to live here.

The last building at the Registan, flouting Islamic teachings against the direct representation of animals that God has made, features a tiled pair of yellow, roaring felines above the door.

"At a distance the animals depicted above the doorway seem to be tigers," Dilshod says. "But they are most certainly not. They are *lions*." The Lion Madrassa, as it is called, is yet another amazing structure, its main prayer hall and classrooms being more of the same, complete with a few merchants. After I walk through this last madrassa, I stand for a long time in the central square in the afternoon sun, staring. Birds flit in and out of the trees. People move across the public space, dwarfed by these structures. Islam may have erected these buildings, but commerce is what created the place. For a time, I walk over and visit with a group of men who've arrived and are playing large, ten-foot-long brass horns on the front portico of the Ulugh Beg Madrassa. As a showstopper, the group's central horn player, the apparent leader, turns his face to the sky and balances the horn vertically on his lips. He then removes his hands and lets the horn point skyward, balancing it with just his mouth and movements of his head.

After that, I spend a few more minutes standing and looking at the Registan square. This place is staggering in its scale. Dilshod breaks the silence. "So," he says, "would you like a glass of tea? There is a good shop, just across the street."

We walk to a café on the far side of the wide Registanskaya. Still, because of the size of the Registan square, it takes us several minutes to get there. After ordering tea and some *samsas*—bigger versions of India's triangular, pastry-covered potato dumplings called samosas—we settle in and visit for a while.

The tea arrives, quickly followed by the samsas, each of which has been cut in half. Beyond the potatoes and peas in their interiors, there is also a yellow curry. They taste like northern India. As I say, thanks to its proximity to the center of the Silk Road, external cultural influences wait everywhere in Samarkand.

The samsas are just right: warmly filling. As we eat, I learn a little of Dilshod's life. Born and educated here ("My father is a religious leader and a teacher," he says), Dilshod has been in tourism now for seven years. "I went to the local university to learn several languages," he says. "I would be happy to speak German with you. Can you speak in German?"

Because of the global economic situation over the last few years, however, Dilshod says, tourism has become a diminishing line of work. I try to shift him away from talk of work. We sip more tea and eat. I show him photos of my family: my children standing in a meadow behind our house; my wife with our chocolate Labrador retriever on our side porch.

Then, finally, Dilshod gets down to business.

"So, as I pointed out before," he says, "like for everywhere in the world right now, times are difficult in Samarkand." He pauses and smiles slightly. He eats a bit more. "My father has been sick for almost a year. He has not worked. I still live at home. I am 28 . . . soon to be 29. I have a woman I want to marry. But, you know, we cannot do this until things are better. I tell you this because . . ."

"I know why," I say. "You were right. I needed a guide, and I learned a lot from you today. For that, I pay you well. But I don't want you thinking that, every day, you will be paid like this."

It hasn't been hard to like Dilshod. He's cool and I have $300 budgeted for guide services while here.

Plus, as a Tajik, he's family.

Pulling out my wallet, I give him the equivalent of about $100—far better than a week's wages. I slide it across the table to him.

"Oh, sir, that is too much."

"No," I say. "It's okay. And please give me your telephone number. That way, over the next few days, if I need some assistance or translation, you can help."

Dilshod smiles again. "This is a deal," he says.

That began my visit to Samarkand. The next morning at 8:15, after a comfortable and quiet night in the hotel, I find Dilshod waiting for me outside the still deserted lobby. He spots me through the glassed front doors. He waves and shouts: "Sir, good morning! They will not let me *inside*."

Outside, beneath the hotel's carriageway overhang, its columns sheathed in what looks like brown marble, I approach the eager young man. "It's okay, Dilshod," I say. "As I told you yesterday, I will telephone you when I need you. You can relax."

"Yes, sir," he says. His eyes are almost black, and a little almond-shaped. He smiles slightly.

"Really, I'll telephone when I need you."

"Okay." He begins to walk backwards down the driveway, back out into the sun, his mop of black hair shining. The doormen smile.

Then I go inside to breakfast. And in that slightly poignant way things sometimes get when East and West collide, the meal proves virtually inedible. While I'm sure the restaurant's chefs are great at making local breakfasts, with regard to making Western-style food they still have a few things to learn.

Inside, other than the manager and the waitstaff, I am the only person in the room. Though not generally squeamish about such things (I've been known to eat fried grasshoppers for breakfast if that's what's on the menu), my fried eggs are already on a white china plate at my table when I arrive, and they appear to have been cooked some time ago. They look like plastic. The honey in a glass jar for my toast, which is also cold, appears to have separated into layers of sugar and extract days, or even weeks, ago. The pulp of the orange juice in my glass has risen to the top in a stopper-like layer. When I ask for a glass of milk, it comes out cold from the kitchen, but it has that overpasteurized taste of having been contained in a Tetra Pak. Even the coffee feels thin. There is, however, a bowl of yoghurt and some muesli on the table, so I don't go hungry.

Still, throughout my 20 minutes in the restaurant, the staff watches me like people monitoring a zoo animal. Together, we get through it.

"Tomorrow," I tell them on my way out, "just yoghurt and muesli and coffee will be fine."

Then it's time to see more of Samarkand. There is the Shahr-I-Zindah, an avenue of exquisitely tiled tombs that might even include the grave of Qusam ibn-Abbas, a cousin to the Prophet Muhammad and the man who brought Islam to Central Asia. There is also the Ulugh Beg Observatory, which Golbor had pointed out to me as we entered the city. It was built in the 1420s to house tools that tracked stars in the heavens. Today, it only holds the footprints of those instruments. And nearby, atop a hill in a small and very pleasantly watered and vegetated valley, is the stone Tomb of Daniel, the Old Testament prophet, whose body is said to be contained inside a 40-something-foot sarcophagus. Legend says that the container for Daniel's body needs to be so large because, even in death, his bones still grow at least half an inch every year.

After my visit to the tomb, as I walk back into the sun-shot afternoon, my taxi waiting for me at the entrance, a gate overgrown with vines, I run into the caretaker. He waves in my direction, then walks over.

"Just over there," he says, "not at all far as a walk, there is a cave. If you would like, for a small extra fee, I will show you this cave."

"Why would I want to see this cave? And how much will it cost?"

"It would be only 2,000 sum," the caretaker says (again, about $2). He points off, presumably in the direction of the cave. "*What it is* . . . well, what it *is* . . . it is the *lion's den,*" he continues. "You know, from the story of Daniel? When Daniel pulled the thorn from the lion's paw, and became the lion's friend? Then, during the sacrifices of Christians at the Roman Coliseum, this same lion remembered Daniel and his kindness in removing the thorn? And so the lion did not eat Daniel with the other Christians? Well, over there is the cave where this famous lion once *lived*."

Apparently some visitors to Samarkand were born yesterday.

Tourism scams aside, what I came to find most fun about Samarkand and the Tajiks were the city's markets. Given Samarkand and the Tajiks' place in the world as traders along the Silk Road, this only makes sense.

The best market, situated in the Old City just a few blocks beyond the Registan, is the Siab Bazaar. Arranged on several levels, and sprawling across a dozen or more blocks, this bazaar crosses (and often chokes closed) major streets. Interested shoppers can purchase anything from a chamber pot to live ducks to lightbulbs and car radiators and plastic squirt guns to the region's enormous spectrum of food. The trick lies in knowing where to look. So each day, about lunchtime, fending off hunger from my lack of a substantial breakfast, I head to the Siab Bazaar for a pita bread and rotisserie-cooked beef or chicken shawarma, with plenty of yoghurt and fresh tomatoes and a Coke.

Then, once lunch is through, I spend the afternoon mixing with the locals.

At the Siab Bazaar, in the open air, sides of beef and butchered goats and chickens are publicly sold. I am offered free samples of dried fish, fresh fruit, candies, and nuts (especially the local delicacy of flame-dried apricot pits, which taste surprisingly like outsize pistachios), as well as yoghurt, and the sweet juice of fresh-crushed blackberries. Two entire warehouses are devoted to the sale of teas and spices. Eventually, I settle into one of the bazaar's cafés, to sip a beer or a cup of green tea and eat another samsa while quickly learning to avoid the *plov,* a meat-and-veggie pilaf that, every time I try it, lies in my stomach afterward like a mound of wet sand.

Still, at the markets, I'm trapped inside my own language, unable to speak Tajik or Farsi. So after a first day during which the local people generously interact by treating me like an overgrown five-year-old, offering items with gentle pushes and smiles and small words and writing down the costs, I ask Dilshod to meet me the following afternoon at the Bibi-Khanym Mosque, which stands near the bazaar, so I can go back to the market to make more sense of it.

A crumbling, domed building with a stately walled courtyard, the Bibi-Khanym Mosque was named for Timur's Chinese-born wife, and was the so-called Jewel of Timur's Empire. After we take a look at the historic building, we walk over to the bazaar together, to snack and haggle.

There are soaps and detergents with names like, I kid you not, Fart and Barf. When we tire of browsing, we eat more samsa, then pause at another table to devour some huge crayfish, each ten inches or a foot long, which have been tastily prepared in a savory but spicy boil.

We wander away from the food stalls, and into an area selling goods out of all sorts of strange boxes filled with different nuts and rocks. This is where the herds-people come to buy dyes for carpetmaking. Dilshod leads me along, pointing out the ingredients as we go. There is "spicy soap": large, pale crystals of aluminum sulfate, which occur naturally and are commonly called alum. While these crystals *can* be used as soap because of their alkalinity, when they are dissolved in water they serve as mordants, leaving wool dipped into their solution more accepting of dyes. Also offered for sale are crushed walnut husks for browns; pomegranate skins for reds and yellows and blues (a broad spectrum that changes with the addition of other different leaves and vegetables and roots); the root of the madder plant, which creates six different pastel pigments (including pink, rose, apricot, and scarlet); then lapis and dried indigo for other blues, mustard for more yellows, and, if you ask around enough, maybe even varieties of crushed beetles for "insect dyes," whose spectrum spreads from lilac to oxblood.

At the bazaar, however, my favorite experiences come from trading with the older peasant women, who are often deadpan funny and who, in their babushkas and frontier dresses with squared gray aprons, seem teleported into today from a hundred years earlier. In the yoghurt building or the spice and tea warehouses, the old women offer me their goods, and when I buy them, they put the yoghurt in plastic bags or fold together small envelopes constructed on the spot from scraps of newspapers. Inside the envelopes go caches of saffron, dried coriander, black cumin, or anise seeds. For fun, I even buy a little "spicy soap."

The best part, though, is this: The women of Samarkand favor gold caps to cover their dental work and tooth decay, and I enjoy cajoling them to drop their self-consciousness and smile and laugh—flashing their gilded mouths—making me smile every time I succeed.

Dilshod and I keep walking, stopping occasionally to buy *shashluyk,* a meat or chicken shish kebab, or circles of naan bread.

"Dilshod? What does this do?" I ask any time I see something that looks tasty or interesting.

Earnestly, Dilshod then asks the saleswoman in the local tongue. There is a minute of negotiation, hands moving back and forth like people dealing three-card monte. Then Dilshod makes a joke and the saleswoman and Dilshod both smile and laugh, the saleswoman's teeth glinting.

Eventually, Dilshod turns to me. "She says this is a leaf off a local tree that works best for headaches when mixed with tea," he says.

"My wife gets headaches sometimes. Maybe I should buy some, take it home and let her try it."

There is more negotiating. More hands moving. Finally, Dilshod turns back to me. "She says it works best with green tea. Tea from here. She can sell you the tea, too. She says you should buy some . . . she will give you a good deal on a bag. And she has a friend in the yoghurt building that she wants you to meet. She thinks you are too thin. She says you need yoghurt from this friend, yoghurt that has been fortified with the fruit of the date palm. Her friend can help with that. You are too thin, she says. She keeps saying that."

It's easy to spend several hours in the afternoon at the bazaar. I can even call it research. After all, starting something like 6,000 years ago, the rise of trade is what ultimately built this city.

As a souvenir, I buy a small, black linen skullcap for my son. It's what all the local men wear. Another day, I buy a glazed crockery chamber pot. In my room at the hotel, there are now several piles of different teas and spices in small folded-newspaper pouches. I eat lots of free samples of yoghurt, much of which is sold plain, with different flavorings sometimes added by the women selling it. Some is bitter. Some is sweet and cool and thick as ice cream. They ask if I'd like honey, or berries . . .

or a kind of small green-leafed plant that looks something like clover and tastes inexplicably *green*. Another refreshing variety of yoghurt, called *chuvot,* has dill mixed into it.

When I get Dilshod to tell the merchant women how wonderful their golden teeth are, the women invariably put a hand over their mouth. Then they hold their other hand open and out, at the end of an extended arm, directing me to look away in their embarrassment.

But during my time in Samarkand, something else is going on, too. The longer I stay, the more alone I feel.

It's a peculiar experience. Other than a few groups of Germans staying at the other big tourist hotel in town—people who travel in mobbed busloads and who I see every afternoon at the bazaars—there are no Westerners anywhere. I remain the only guest at the opulent, atrium-lobbied Hotel President. And as if to reinforce my *Little Prince*–like segregation, each time the hotel staff sees me walking in or taking a taxi up the driveway in the late afternoons, they flip on all the lights inside the otherwise dark and soaring atrium. It is no fun to sit in a comfortable chair in the lobby—all by myself. The hotel restaurant remains deserted, and the empty lobby bar appears to have never poured a drink.

So, instead, from morning until after nightfall, I wander Samarkand's sidewalks, avoiding the prohibitive cost of a phone line to the United States and, for some reason, unable to secure a steady connection to my AOL account at any of the city's numerous Internet cafés. Wimbledon is happening outside London, though the tennis matches come in only intermittently via satellite TV in my room; the picture keeps breaking into hours-long blizzards of digitized snow. Over the next few days, I see every landmark in town, interacting with some of the locals in tea shops and bars. And though I remain a fan of both Samarkand and beer, the combination somehow doesn't leave me enjoying late afternoon beers in Samarkand very much.

It takes only a few days, but soon I am bored . . . mostly with my own company. Somehow, I've become trapped in a solitary, modern limbo that's left me both jumpy and a little uneasy.

Soon I am reminiscing about peacefully pleasant days spent with Julius and the Hadzabe, where life's pace makes sense, overlaid by the simple, cause-and-effect realities of finding food in a relatively stable environment. Evenings filled with the peace found inside the ruins at Baalbek also come to mind. In the past few weeks, I've gone from being a hunter-gatherer, cooking by campfire in a world where events led naturally to other events, into a life spent stumbling through a maze of specialized and sputtering social constructs and communications devices. My leap across 25,000 to 30,000 years of human progress has left me disoriented. My brain is tired and confused, and—at the same time—spinning at capacity in a vacuum of limited input.

Of course, it could be worse. Unlike that other American who deplaned with me back at the airport, at least I haven't had all my worldly goods whisked into that blind infinity called Lost Baggage.

Still, my mind is having a hard time processing it all. Late one afternoon toward the end of my visit in Samarkand, standing on the sidewalk beneath the spreading and shady trees of the ancient Silk Road—halfway around the world from my home—I am frustrated and alone and lost, invisible in the bigger world.

And that's when it hits me. In a very real way, this is what all of us are doing in modern life: running cutting-edge, sociocultural software on computers called human brains that haven't been truly upgraded in *at least* 100,000 years. It suddenly seems little wonder that sometimes in the news we see stories of people who plummet into inconsolable panic and despair, or choose to disappear, or go violently and incomprehensively berserk, their lives derailed by a confluence of gray matter short-circuits and avalanches of toweringly confusing external stimuli.

Across this trip, in going to meet my extended family, I have willfully walked away from my known world, relying on telephones and e-mail and TV to keep me lightly tethered to the comforts of my usual life. Now I've ended up halfway around the world from home, far, far out onto the conceptual scaffolding of human connections, with parts of that scaffold having collapsed behind me. I have deposited myself on a uniquely modern form of island.

Hell, I can't even speak the language.

Standing in the sun on this beautiful day, it becomes obvious that, all at once, the tools of modern life—and the belief systems underlying them—have abandoned me. The tools are taking their revenge.

---

What saves me is a circus.

On my second to last day visiting Samarkand, riding in a taxi and searching the outlying districts of town, I spot the colored panels of a big top towering at the end of a wide, cul-de-sac. In the heat of Samarkand's midday summer sun, a raggedly dressed man in overalls and without a shirt on sits atop the tent's roofline, 80 or so feet above the ground, rinsing the structure with a long black hose he's holding in his right hand.

"Wait here," I tell the driver.

Before long, I wander inside the empty big top. The air is hot and doesn't move. It's stifling. And then, in walks Tugol, director and ringmaster of Circus Tashkent, the last traveling circus in Uzbekistan. At the moment, Tugol—a pale, thin man with black hair parted on the left—stands inside the tent, dressed only in athletic shorts, a gray sleeveless T-shirt, and blue plastic sandals. Yet he seems all too happy to talk, and to show me around.

"This is Sunday, always our cleanup day, so excuse our appearance," he says. "But, still, we always must do our cleaning, to get ready for tonight's show: our Seven O'Clock Spectacular."

In minutes, my arrival has spurred people into action. Other Circus Tashkent performers have stepped from their chores and assembled, still dressed in their cleanup clothes, hoping to exhibit their skills. The first are two women, one of them breast-feeding a newborn, who say they'd love to show me their high-wire and trapeze act. But, pointing at the cable sagging overhead, they say the day's scrubbing and triple-checking of their rigging will unfortunately keep them grounded.

There is also Alexi Loha, a shirtless blond 18-year-old in denim overalls, who slipped away shortly after introducing himself and returned with a rhesus monkey sitting on his shoulder. Then he launches into his and the monkey's act; the monkey tumbling, doing backflips, and generally mugging around as Alexi gives it instructional cues and treats.

Then there is Bowa, the circus's dwarf clown. He's dressed in trousers and a T-shirt, and—after Alexi launches into his show with the monkey—he walks off a distance under the big top. When he gets to the circus tent's door and notices I am still watching him retreat, he flips over and exits the tent, walking out on his hands.

A few minutes later, Bowa returns right-side-up with a 12-foot python. It's as thick as a fire hose with diamonds of black scales running down the spine of its brown body, and it's slung across his shoulders. There is an incandescent smile on Bowa's face. The python moves slowly, its body gently writhing, its black and brown scales sliding over one another beautifully inside the strange, color-wheel light of an orange and yellow and blue circus tent beneath the midday sun.

As I relax with the clowns and tumblers and acrobats during their time off, I begin looking around, checking out this place. The inventory isn't pretty. The seats are row upon row of battered wood bleachers, their paint chipped and worn. The tent's raised center ring is patched and peeling, its sections only latching together loosely. Beneath the ring, the cul-de-sac street's concrete is covered by a dingy and stained stretch of green indoor/outdoor carpet, meant to imitate grass.

For a time, I talk with Alexi. He comes from the farming outlands, was trained for the circus in Moscow and Tashkent, and has now been with Circus Tashkent for five years. He enjoys it, and says the performances and the slightly itinerant life are fun. "But then, I am still young," he says. "So it is all mostly new for me."

I tease and play with the monkey on Alexi's shoulder, then chum up with Bowa, who loops the python over his *and* my shoulders and asks another of the Circus Tashkent performers, a taciturn woman who is working halfway up the ladder leading to the high trapeze platform, to take a photo using my camera. She climbs down and snaps a few shots.

"I have an e-mail address," Bowa says just after our team portrait is taken. "I will write it down for you. You will send me this photo, okay?"

Still, no matter how welcoming and genuinely nice the workers of Circus Tashkent are, the scene is more than a little depressing. As I sit there, in the hot shade of the multicolored tent, Tugol notices me giving his battered empire a hard-eyed once-over.

"You should come to the performance tonight," he says. "You wouldn't believe how different this place looks at night, when it is lit by spotlights instead of the light of day. I will get you a *free ticket* for tonight. It is a show worth seeing. A circus is really theater, after all."

Recent years have been hard on Russia's former circuses, Tugol says. "At the height of the Soviet Union, in the late 1980s, there were 70 permanent circuses and 50 traveling circuses across the republic," he tells me. "Then everything collapsed. Though people need to see a circus more than ever these days to lift their spirits, fewer people come. But we keep going. We can no longer afford horses; we had to get rid of many of our large animals. They were too expensive."

Now, Tugol notes, Circus Tashkent is down to a monkey, a snake, a bear, and a donkey that has been painted in black and white stripes, to look like a zebra. "If you would like, I will introduce you to the animals," Tugol says.

"Sure," I say, though I've already met the monkey and the snake. Still, the tour will get me out of the crushing heat inside this tent. "But first," I add, "I have a question. If Circus Tashkent is so hard to keep going, why do you carry on?"

Tugol smiles broadly. He gives a little wave to the central ring on the floor. "Because the circus is what I *do,*" he says. "And I love it. Each night, if no one comes, I still do the performance to an empty tent. To be a ringmaster, lit by spotlights, with performers in sparkling costumes, it is me saying to the universe: We humans are *amazing*. Look at what we can *do*. We have trained these animals. We have bright lights. We have breath-taking human skill on display. Look at us: *We are a circus!*"

Yeah, I thought. He's got it exactly.

The last night in town, I eat at an outdoor café about a 15-minute walk from the hotel. A small, rocky creek tumbles through an area between the restaurant building itself and the gravel plot where the outdoor tables sit. A wooden footbridge connects the two. There are candles glowing on all the tables. The sky stays sunset blue and orange for a long time, and overhead, the trees spread their limbs over the scene. A small string band plays traditional songs, some upbeat and others slow ballads. Once in awhile, the guy playing a guitarlike lute sings in what I can only guess is Farsi.

At the table, a tomato salad and French fries show up first. Then grilled mutton rib steaks and the veggie pilaf called plov. I order a glass of local red wine. Thanks to rising worldwide trade in the last few thousand years, the ancient ancestors of these fried potatoes showed up from their own species' birthplace in the Peruvian Andes 400 or so years ago. The salad, mutton, plov, and wine have probably been cultivated somewhere near this place, but their ingredients' birthplaces probably did not originate near here, either. Now, trade has brought them all to Samarkand. In a way, the history of the modern world sits on plates in front of me.

Then comes vanilla ice cream for dessert. Ice cream is an invention of Europe, though the seeds from pods that flavor it originally came from orchids in what is today Mexico.

After dinner, I walk back to the hotel, happy for the wine and the mutton and the beautiful continental collision of history between vanilla beans and ice-chilled dairy cream.

As I approach the Hotel President, Dilshod is there, standing in front of the hotel beneath the covered carriageway.

"Sir," he says as I approach, "they would not let me into the hotel *again*." He flashes his usual, slightly embarrassed smile. "But I wanted to say goodbye."

Of course, aside from wishing me bon voyage, there's likely another reason Dilshod is here. I've spent only about $100 of the $300 budgeted for guides on this section of the trip, and he's probably hoping for a tip . . . not that I can blame him. As I walk up the driveway, he smiles again.

While never averse to saving a little money, I've also come to like and sympathize with Dilshod. Judging by what he's told me, on the days when he's not been with me, he seems to spend a lot of time sitting around and worrying about his future. Though I decide to keep some of the budgeted money back, for tips at the hotel here and the driver tomorrow, and then for tips in Tashkent and at the airport for my trip out in two days, Dilshod does deserve something more. I pull out my wallet, unburdening it of much of the Uzbek sum I still possess.

"You've been great," I say. "Please think about putting some of this money in a place to save toward your marriage."

Then I hand him the cash.

In the spot-lit darkness of the carriageway, Dilshod smiles and nods. "Yes, yes sir," he says. "Would you like me to see you off tomorrow? To talk with the driver for you? To give any special instructions?"

"No, that's okay. The hotel got me a car. The manager can speak English. It's taken care of."

"Yes, sir."

We shake hands. "Thank you," I say.

"Thank you, sir."

Then I begin walking into the hotel. In the morning, I'll be headed northeast, up the Silk Road toward Tashkent, with China sprawling out there somewhere beyond. Now, in the evening, Dilshod walks down the driveway, and disappears along the Silk Road in the other direction.

On my last morning in Samarkand, something truly unexpected happens. As I wake up and realize it's my last day in town, I have to admit to being a little disappointed by the visit. Though I've enjoyed several aspects of Samarkand, I've not felt the same connection with the Tajiks that I did with the Hadzabe in Tanzania or the Lebanese Arabs of Baalbek. Maybe it's cultural. Or the fact that I can't speak the language. Maybe it's that this place, with its ancient focus on commerce more than companionship, just doesn't evince warmth to strangers. Maybe it's that, having a hotel to myself and being detached from my usual means of communication with the English-speaking world, *I'm* the one who has become alienated, and everyone else is fine?

Whatever the reason, though, there's been less of a sense of connection. Which is too bad, since—despite the otherness of this place—I truly like Samarkand.

A while later, as I come to the lobby to stare one last time at my breakfast and consider what I'll soon eat from a street vendor in advance of my ride at 10 a.m., the desk manager, a medium-size guy with dark eyes, a long, straight nose, and a black comb-over, hails me from across the hotel's vast, unpeopled lobby.

"Sir," the manager almost shouts. "We have something here . . . for *you!*"

I walk to the front desk as the manager bends beneath the counter, his hands reemerging with a parcel wrapped in creamy white tissue. "This arrived late last night," he says. "The night staff assumed you would be asleep but would come down for breakfast, so we waited. I hope this is all right."

He places the parcel on the counter between us. It's about the size of a book, but appears not to be rigid.

"Of course, of course," I say. "Thank you." Then I grab up the package and untie the string binding it together. The layers of tissue crackle as they peel back.

Inside, wrapped in a bag of unwrinkled cellophane, is a beautiful silk shawl: in a thousand shades of purple. As I lift it, the colors actually shift and change in spectral grades, even in the murky, flat light of the hotel lobby. Out in the sun, or in any natural light, this thing will be gorgeous.

With the shawl is a small white envelope. Inside the envelope is a stiff white card maybe three inches across. On the card, inscribed in black ink, are these words:

*Sir,*

*This is my Thank You for the time we spent together, and also for your generosity to me and my family. As a gift to you and a reminder of Samarqand and the Silk Road, it is my hope that you will take this drape for your wife. It will make her even more beautiful than she is in the picture with your dog. When she wears it, I hope you will think of me and Samarqand.*

*Your always friend,*

*Dilshod*

*A Basque girl in San Sebastián, in northwest Spain—yet another distant family member in a world of shared genetic history*

# SPAIN

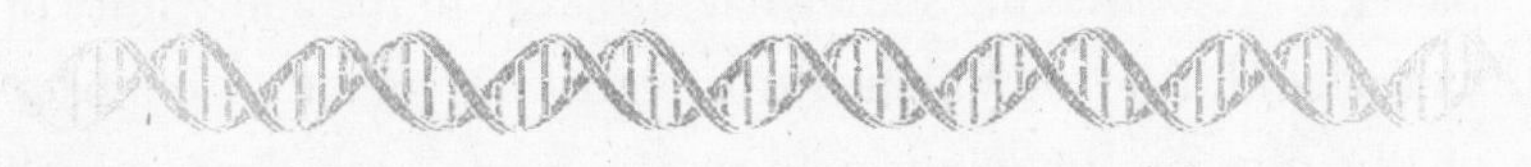

**THEN WITH A SPEED** that feels like switching channels on a TV, I am in Bilbao.

Just as in Beirut, my hotel has sent a shiny black BMW to collect me at the airport and carry me into town. Outside, the weather is classic maritime Spain. As if divided by huge and different-colored plates, the western half of the sky, which extends out over the Atlantic, is gray and spitting rain onto the rolling emerald green and mountainous landscape; a place so verdant that even the rock-chunked hills appear enticingly and edibly tender. The other half of the sky, the inland half, is sunny and spotted with just a few puffy clouds that leave huge shadows moving across the landscape's roll. Even the sky here seems to have volume. To the north, a line of Pyrenees foothills rises up, equally green and something like a wall.

Bilbao is my itinerary's final stop: the Basque country of northwest Spain. And after five weeks of travel, meeting people

I share genetic markers with and taking notes and learning and trying to soak up the experiences of these very different places, I'm exhausted. I've been hunting and gathering and sleeping on the ground in Tanzania, going through the heat of the days at Baalbek . . . wandering somewhat aimlessly in the anonymity of Samarkand. And now there's this place, and seated in the back of this nice car, its windshield wipers going each time we roll through a small rain shower, it feels as if there's no more room for thinking. No more energy for note taking. No more curiosity to seek new people out. Mentally, I'm wrung dry.

Still, the landscape is gorgeous. And I can speak the language of Spain, which at least will leave me less isolated. The stay here is scheduled for a bit less than a week, and that buoys my spirits, too. A week where I can communicate with people, where I have a nice spot to sleep, and have an interesting city to explore? That's doable. Maybe it'll even serve to recharge the quest.

So as I sit in the back of the car, I think of Spencer Wells, and one of the last things his said to me before the trip.

"The people you descended from, haplogroup *R1b,* gave rise to the first humans to move into Europe and eventually colonize the continent," he said. "Because of markers *M173* and *M343,* you're a descendant of the Cro-Magnon; they created the famous cave paintings of southern France and northern Spain. But yours show a tendency toward Basque Spain. You're hardly alone in that. A large majority of the men in southern England and northern France share these same markers. We believe that people with this marker were living in Spain and other areas of southern Europe until the glacial ice sheets of the last ice age began retreating, about 10,000 years ago. As the ice moved back north, people followed it. And that's your heritage."

With that, the story of how my people got to England and beyond is completed; I already know the rest in terms of migrations to America and the relentless advancement of life leading to the modern-day Miracle at 36,000 Feet, jet flight.

In fact, for the last 10,000 years, for all of us *Homo sapiens,* it's just been more of the same: each generation making the world more comfortable and usable through a process of trial and error, quite regularly with astounding success. For the last 10,000 years, we've been developing ideas and running with them, over and over, until we've either refined them further or they no longer apply. Think of the ascent and decline of the ancient Romans. Or the way the Spanish conquistadores and the Carolinian English and the Hudson's Bay Company colonized the New World only to be rendered eventually irrelevant by those they once directed. Or that brashly confident Emperor Napoleon, who in 1812, marched into Russia with 449,000 soldiers, having defeated or cowed most of Europe. Then, arriving in Moscow to find the place burned to the ground he and his army faced a long, wintry, bullet-ridden walk back to Paris that will reputedly leave Napoleon with only about 22,000 men. Recall, too: "The Sun Never Sets on the British Empire . . . ", the faceless, meat-grinder killing across the poppy fields of the Somme. . . the Thousand-Year Reich . . . nuclear power . . . the rise of silicon-based memory . . . the notion that the Internet will save us by connecting us. . . .

As *Homo sapiens,* we move relentlessly forward, always leaning more aggressively into our discoveries and what they can do for us: creating yet another generation that crashes into the new obstacles our advancing prowess places in our way. That is, unless our "advances" crush us first.

In the world *Homo sapiens* has made, this is now the process that goes on and on, over and over. Still, as a journey, it's something to marvel over: a long road from thrown rocks and Acheulean tools to painted cave walls to today.

Like many of the rest of my extended family worldwide, the Basques remain uniquely and fiercely independent. With a population locus in the hills of northern Spain, spreading across the Pyrenees into southwest France, the Basques have been lodged in this part of the world so long, they like to point out, they have no migration or arrival myths in their oral history.

Deeply agrarian people with a long social history, their traditions are nothing if not idiosyncratic, and they include whole fistfuls of unique social, sporting, and sartorial customs. These include gruff greetings followed by warm hospitality and a love of quirky sports that range from jai alai to hand-scythe grass-cutting to bullfighting and bull-running. Then there is the area of dress: those horizontally striped tunics the more rural Basques still wear with neither irony nor self-consciousness.

There's also the native tongue, whose origins lack any linkage to other families of speech, confounding linguists. And jealously guarded is the Basque flag, the *Ikurriña,* a kind of different-colored Union Jack in red and green. It's an omnipresent symbol of the Basque Autonomous Region, or Comunidad Autonoma

Vasca. Never far out of sight, the Ikurriña reminds the locals of their proud history, not to mention more than half a century of attempts to destroy local Spanish and French rule in the name of cultural independence.

This brings us to another, somewhat notorious, Basque tradition. Inside what is said to be a small yet radical group, there's something called Euskadi Ta Askatasuna, or ETA. It's a sometimes-violent extremist organization that sponsors regular kidnappings and public bombings throughout the region . . . and beyond. Translated into English, the name means Basque Homeland and Freedom. And they mean it. Since basically the time of Generalissimo Francisco Franco, who attempted to force the Basques into Spain's otherwise tidy national government after World War II, ETA has been fighting back, often with bombings that destroy power plants or government facilities, usually timed to take as few casualties as possible.

On December 30, 2006, for example, ETA took credit for the bombing of Madrid's Barajas Airport, killing two men who were napping in their car in a parking area. In February 2008, they blew up a TV transmitter station on the outskirts of Bilbao with no fatalities. In July 2009, outside the Spanish Socialist Workers Party building in the city of Durango, a bomb caused extensive damage to the building, but no one was hurt. Every year, ETA takes claim for dozens of bombings and kidnappings. The toll of dead rarely climbs into double digits, but a low casualty rate leaves such behavior no more acceptable.

Arriving as a visitor hoping to burrow beneath the exterior of ETA's home, I find this lowered death toll of slight comfort. Still, as ETA was formed as a student organization in 1952 in Bilbao, the very city I'm rolling into via a shiny hotel limousine, I'm steeling myself for the possibility that, well, during my roughly weeklong stay, something nearby might go *boom.*

Yet the presence of ETA and its violent slashes toward Basque freedom don't really get close to encompassing the breadth of the Basque spirit.

Better to stop by the Café Boulevard, along the promenade by the Bilbao River at the edge of the old city, and let Enrique Cardenas take control of your late afternoon. The café is the oldest in the city, it dates from 1871, and with absinthe green art deco paint and mirrors covering every wall and column surface, the place looks as if Hemingway and his "Lost Generation" buddies might be seated at any table inside.

In the center of the café, behind the bar, stands Enrique. Sturdily built and dark, with craggy facial features seemingly cut from stone, Enrique appears off-putting the first time I order a glass of red wine and two small plates of tapas (or *pintxos* in Basque parlance), then unfold a crisp *International Herald Tribune* across the bar's wooden top.

With his stern schoolmaster's face, he brandishes the green bottle of wine and pours; then he places the two open-faced sandwiches I've ordered, one of Serrano ham and Manchego cheese, the other of white tuna with olive oil and pimentos, on a small white plate and slides it in front of me. A few minutes later, after I have enjoyed both bites and finished the wine, Enrique returns.

"Thank you," I say. "That was very good." He smiles. And, suddenly, it's as though I've run into a boyhood friend. "No, no," he says. "*Now* you must have *another* glass of wine. This one I'll pay for. It is a favorite, from a region in the district of La Rioja south of here. A beautiful place. They make very good wine. You must also try this: fresh anchovy pintxo. Also this, with whole baby eels and olive oil on the pimento and the bread. And after that, I'll get a few hot plates: fresh shrimp, just off of the boats, with garlic and olive oil, and perhaps another plate of fresh mackerel."

Before long, I've told Enrique of my quest, and he and I are deep in conversation about what it means to be Basque, a clan of which he is an admittedly proud member.

Or, more accurately, for the next hour, I listen to Enrique talk about what it means to be Basque, as he guides me through the highlights of the café's menu. "To walk along this ancient river," he says, "to walk our green mountainsides in the sun . . . to enjoy the food and wine, just as my grandfather's grandfather did in this place. This is our history and identity. . . . This is the source of love for our home. . . . This is what it means to be one of us. . . . This is what it *means.*"

Something else it means, however, carries darker portents. Just as it does for anyone living today—or who has lived during the last 6,000 or so years.

Starting about then, with the rise of regional trade and the languages and eventual Bronze Age technologies this trade carried with it; we entered the Age of Technology Traps. Waterborne salt introduced into the soil of Mesopotamia through the early success of irrigation was among the first of these dead ends.

But that just means the Mesopotamian technology trap came early on. The ancient Egyptians, united under the first pharaoh, Menes, about 5,100 years ago, rose to become their own highly stratified agricultural society, only to see it break down three times into squabbling principalities run by leaders interested in exploiting those beneath them to advance their own legacies through taxes and architecture. By the time another cultural flowering happened along the Nile in ancient Egypt, about 332 B.C., when the late Egyptians began to meld with the ascendant Hellenistic culture across the Mediterranean, even the middle class and more successful farmers of Egypt were gilding their funeral wraps, spending current resources for eternity.

Taking their cues from their society's royalty, Egyptians were abandoning their time on Earth to concentrate on the afterlife. Perhaps the ultimate technology trap.

Yet it didn't end there. The ancient Greeks even saw the end of their reign coming, and couldn't do anything about it. Thanks to the rise of their populations and the diminishing returns of local forests and agriculture, plus the slow exhaustion of their earth through overfarming—and a growing serfdom where the powerful manipulated the weaker—they eventually contributed to the destruction of a too-fast-growing republic.

Around 590 B.C., the Greek statesman Solon suggested an end to debt-serfdom, as nobles overseeing farmland had promoted the growing of crops on Greece's arid and not-terribly-fertile soils for personal enrichment until that ground produced little more than rural poverty. Even as Solon pointed this out, the political power of the Greek *poelsi* weren't up to the task of agricultural reform. Two centuries later, in his dialogue titled *Critias,* the philosopher Plato wrote this of the crumbled landscape that survived and of the society presiding over it: "What now remains compared with what then existed is like the skeleton of a sick man, and the fat and soft earth having wasted away, and only the bare framework left. . . . Now only abandoned shrines remain to show where those springs once flowed."

The dominion of the Greeks over the "civilized" world was followed by that of the ancient Romans, who took another direction, and one that led into a separate technology trap of their own design. Across four centuries, they grew their empire from a networked republic of peoples into a more centralized, power-based government in Rome. During the almost 500 years of their control of Europe, the ancient Romans took more and more lands and peoples inside their empire, while still trading with outsiders in an increasingly complex and multilateral way. In doing this, they staved off the "exhausted earth" principle by outsourcing the demands of their populace to farther-and-farther-flung farmlands.

Thanks to all these diverse populations and landscapes and external trade, the ancient Roman upper class remained both wealthy and well fed. They did suffer a series of pandemics, however, which taught them the value of sanitation across their growing network of cities. (The ancient Romans were some of the first to employ public bathhouses and toilets, the waste transported away through a system of canals that ran beneath their buildings.)

Still, they couldn't escape corruption and despotic rule, not to mention another problem of their own making. By the time of Emperor Constantine around A.D. 300, keeping the sprawling empire together required an enormous standing army, one said to number far more than 375,000 men. Spread as it was to the edges of the Roman holdings, the expense of feeding, supplying, and maintaining these forces became difficult . . . and then impossible. Roman treasuries couldn't keep up with their armed legions' needs.

At the same time, across the growing outer edge of their empire, the Romans were being beset by more and more outside armies, organized by people looking to keep their freedom, even as those just inside the empire's lines were paying taxes to a ruler who seemed less than invested in their well-being. It didn't take long before some of the outlying Roman citizens and troops, hungry and underappreciated, began to defect and to oppose Rome alongside their former opponents, setting into motion the Germanic and Slavonic migrations through Europe and taking the glory of ancient Rome down with it.

And there are more examples of technology traps, which pile up faster and faster as societies grow more sophisticated. No one is exactly sure what happened to the mound-building cultures of North America, or the ancient Maya in Mesoamerica, both of which, for example, started about 4,000 years ago and thrived in a world of cities and temples and pyramids before simply disbanding. While sediment studies near the Maya Empire's central city of Tikal show erosion lowering crop yields as one underlying problem, some scholars believe disease, peasant uprisings against a growing royal culture, and internal warfare between clans or groups may have contributed, too.

Or consider the Mongolians. Starting in the year 1206, their charismatic leader Ghengis Khan—or Temümjin to the locals—united the Mongolian people of the Asian steppe, then sent them violently into the world on horseback. Using bows, arrows, and sharp blades, and keeping live meat on the hoof with them for nutrition as they traveled (taking more livestock from conquered cities and tribes along the way), they were more successful as attackers than any nation had ever been before.

In just over seven decades the Khans conquered more land than the Romans had in more than 400 years. By the time the Mongolians' assault on Asia was over, their horseback armies had taken most of the continent: from the beaches of the Black Sea in the west to China and the Korean Peninsula in the east. They shattered the Great Wall of China, specifically erected to keep invaders out, installing themselves for the next century as China's emperors. They razed Baghdad. In their assaults, they were systematic and centralized: Ghengis Khan and his two descendants decreed that, no matter how big the army, each encampment along the way was to be arranged in a grid, with the leadership at the center. They also refined the idea of terror as a political tool and the notion of political emissaries: sending someone from the most recently conquered city ahead to the next city, to advise of the carnage that had arrived with the Mongolians and suggest preemptive surrender.

But the all-conquering Mongolians also developed a problem. They'd embraced the idea that conquering lands was the whole point, and they didn't manage them very well once occupied. Their hunger had gotten the best of them. And like the ancient Romans, eventually the scope of the real estate and people they ruled became ungovernable due to its expanse and underlying cost.

Eventually, the people governed by the Mongolians took their freedom back.

Across the Neolithic world, starting about 6,000 years ago with its solidification in farming and villages and established trade, the lesson, over and over, is that increasingly elaborate social systems are usually overwhelmed by their own complexity.

Sometimes the societies managing these systems simply run toward entropy (as happened in the case of imperial Rome), leaving behind ruins of their past in the form of languages and practices and still functioning bridges and aqueducts that extend into future millennia. Other times, as in the case of the Anasazi and Maya, the society effectively disappears, leaving behind pottery sherds or step pyramids whose construction techniques our archaeologists and scientists still don't completely understand; their existence tantalizing evidence that accrued knowledge can suddenly evaporate forever.

And don't think it stops with the invention of gunpowder or the automobile or electric lighting. It's a dubious indictment of human nature that, for instance, we have now grown so good at fighting wars we have whole classes of weapons, poison gas and land mines and lead bullets and nukes, that we've collectively decided should never be used. We have become so good at trawling the seas that a recent newspaper story has been read around the world: The town of Oma, on the island of Honshu in Japan, after surviving on its bluefin tuna harvests for thousands of years, has now effectively been abandoned as a result of the collapse of its fishery, owing to newer and "more efficient" bluefin harvesting technology. This technology has fished Oma out of its future. And, given the way we're harvesting the oceans worldwide, Oma may soon stand as only an early casualty.

By increasing the complexity of our world, with technology creating new belief systems that birth more new technology, the traps we have laid for ourselves wait quietly everywhere, like the land mines we first invented to keep us safe and now struggle to outlaw. In Oma's case, the story is nothing but tragic. Still the people of Oma will pick up and go on. Fortunately, in the case of nuclear weapons and land mines, let's hope our collective "moment of clarity" arrives before the tools take their ultimate revenge.

Once in awhile, we don't need to get all the way to the dead end of an alley before comprehending that there's nowhere left to go. After all, the human learning curve may not always be steep, but it's rarely completely flat, either.

A case study in societal collapse.

It was evening in mid-March of 1999, and I was sitting in an oasis in central Egypt, a place called Bahariya. With me was

Zahi Hawass, now director of all antiquities for Egypt but, at the time, director general of the Giza Pyramids, Saqqara, and much of the central Egyptian nation. On this evening, we were at a *taftish,* an antiquities headquarters, small museum, and warehouse-like storage area for the mounds of ancient Egyptian artifacts being unearthed there.

Hawass and I were in Bahariya because, not much earlier in the week, the modern world had stumbled onto an ancient mystery. A few years earlier, in 1996, a local man whose job it was to guard a tumbledown ancient temple in the desert outside town was riding his donkey to work. Then, about half a mile from the 2,300-year-old temple, the donkey's hoof literally fell *through* the ground.

"The guard got off his donkey," Hawass said. "He helped the animal up, and looked into the hole that had been created. Inside, he saw a tomb."

After thousands of years of people, not to mention donkeys, traveling across this desert about 200 miles southwest of Cairo, a new window into the ancient world had opened. Once alerted to the find, it took Hawass a time to get professional and museum-quality excavation teams in place, but when he did, I was lucky enough to be present. And what he discovered was amazing. The area was filled with tombs, and inside were aggregations of mummies from the Greco-Roman age of ancient Egypt (which began with the arrival of Alexander the Great in Egypt, in 332 B.C.). The upper bodies of many of these mummies were covered with *cartonnages:* rectangular head and chest plates of plaster, which were then covered by layers of gold.

The more Hawass and his excavators kept looking, the more golden mummies they found. "We estimate there may be 10,000 of them here," Hawass said, his piercing eyes smiling as he flashed a leonine grin. As always, he was dressed in his field uniform: light boots, khaki trousers, a red button-down shirt, and an "Indiana Jones" fedora. "We're calling it the Valley of the Golden Mummies," he added.

Testing on these artifacts had already shown the age they came from, Hawass said, a period when ancient Egypt rose again from its ashes and prospered. With external threats quelled through political alliances and international trade, Egypt's internal life was stabilized and entering a new period of prosperity. No longer preoccupied with defense, and with more efficient cultivation as a larger variety of foods was being imported, Egypt's people flourished and began to spend more time enjoying their lives—a luxury that left them thinking and living more like the nobility that, until then, had dictated their existence.

In the days of the pharaohs, Hawass continued, only the ruling elite had been able to afford golden funerary masks, with the Egyptians bartering for most of their goods and services. When the Greeks arrived, coins were more widely introduced into Egyptian culture, which eventually led to a wider distribution of wealth across Egypt. More people could afford to buy burial masks, and workshops began to turn them out. The plaster ones with gilded tops had the look of royal masks without the astronomical cost. As wealth grew, the upper and middle classes began to commission masks, as well. To tell each mask-wearer's story, some artisans were even asked to create the gilded cartonnages whose symbols told of the deceased's place in society.

But even as ancient Egypt flowered once more, the historic reasons behind some of its practices were disappearing. "By the Greco-Roman period, in the ways of mummification, most Egyptians were merely going through the motions," said Nasry Iskander, general director of conservation at the Egyptian Museum in Cairo, "They were carrying on this tradition, but maybe they could no longer remember why."

At Bahariya, instead of mummified bodies being buried facing east, toward the rising sun as a daily symbol of their eternal rebirth, they are buried pointed in all directions. And few were buried in coffins, with several bodies often piled into the same tomb niches, a practice that had never been done previously. Also, of the 105 mummies excavated and examined so far, Iskander told me, none were buried with canopic jars, which contain the internal organs of the deceased for use in the afterlife, items commonly found in tombs predating the Greco-Roman era.

But, as with other cultures, a loosening of traditional religious practices by dilution with outside beliefs, plus growing time pressure on embalmers because of increasing demand, may have changed the way things were done. Still, there were other reasons things were evolving, too. And these influences were even bigger than new cultural and commercial trade.

Like the slowly tiring soils of Mesopotamia, the earth in Bahariya was delivering less each year in the way of bounty. The soil was growing salty there, as well. These days, what's grown around Bahariya is largely forage for animals, as the soil is too tired and depleted for growing more demanding crops. According to Hawass, judging from the symbols found on the gilded mummy cartonnages of the Greco-Roman period, the people of that time were prosperous farmers and overseers of those farms. "We have found evidence that Bahariya was a large government winemaking center," he said.

But something more was threatening change, too. The oasis was slowly drying up. Recent excavations of early aquifers at Bahariya show the water available for drinking, washing, and irrigating sat just beneath the surface. By the Greco-Roman period, judging from excavations, that water lay 15 feet down. Today, wells must be sunk as much as 4,500 feet in this valley to find what water remains.

As we talked on this evening, we were sitting deep inside the taftish, in a room filled with artifact-choked shelves and wooden cartons stacked waist high, which had recently been brought in from the site. We'd spent the last hour looking at some of the gilded mummies. With their curly hair, long aquiline noses, and wide eyes inset with mother-of-pearl, the masks reflected the influences of Greek sculpture. They were far different from the almond-eyed and smoothly gilded headpiece of, say, Tutankhamun. Around these newer mummies, archaeological excavators had also found unusual foot-high clay statuettes of wailing people: Their mouths open in groans of pain, their arms out to the sky. Hawass said he believed these were symbolic mourners, to denote the importance of the person buried. The more statues they were buried with, the higher that person's social status.

Hawass stood then, and walked to a table across the warehouse-like room. From its surface, he lifted a four-inch-tall sculpture of a perched falcon, its exterior glazed in faience and fired to a rich greenish blue. The bird was perfect. Its individual feathers were visible and detailed. Its wings folded to a graceful V behind its back, its hooked beak appeared ready to rip open the next prey in its talons. Emanating both elegance and antiquity, it was one of the most beautiful objects I'd ever seen. "I like this piece; it came from one of the tombs," he said.

As Hawass handed it to me, I couldn't help but wonder who was the last person in ancient times to hold it. Who, like me, regarded this sculpture as so perfect they wanted to take it with them into the afterlife?

But before I could ask, Hawass had taken the sculpture back, and returned it to the tabletop. He then lifted an eight-inch square of limestone etched with concentric circles. "Look at this," he said, handing me the slab. It was heavy as a brick. "It's an early board game, found in the tomb with the gilded mummies. Sort of like backgammon. Can you imagine the ancient Egyptians playing this *very game*?"

The stone's bull's-eye had been scooped out, and a small cube of etched limestone, a die, teetered in it. I held the game gently, careful not to tilt and roll the die. Who made this? And how many days and nights did they spend huddled over it, immersed in contests filled with strategy and thought, teasing and wagering?

"There were not the social diversions here that existed in the cities," Hawass said. "So they played games; had parties. Can't you see them, eating dates and olives? Drinking wine? This is why I love my job so much."

He took the stone back from me then, returning it to the table. He was smiling.

"So why do some societies last, and others disappear?" I asked. "Why did this place prosper, with gilded cartonnages and wine, and then become lost to the world for 2,000 years?"

Hawass grinned again, his eyes sharp. "That is the same old story," he said. "A society rises, becomes stratified and diverse and great, and then, slowly or quickly, it can deteriorate and die. Sometimes it dies a mystery. Sometimes, it stays around in smaller pieces. Often, it takes much of the knowledge and expertise it developed with it to its grave."

At present, Hawass believes modern science has discovered only about 30 percent of the riches of ancient Egypt. And while science and research have made sense of a lot of what has been unearthed from beneath Egypt's sands and pyramids and temples, as time goes on we will come to have a better understanding of ancient Egypt and what it knew and what it stood for.

"But right now," he said, "I estimate we know only about 30 percent."

He believes he may have finally teased out, for example, how the great stones that make up the Pyramids at Giza were quarried and cut so cleanly, then transported to the pyramid site, several miles away. "It's my understanding of it that a single man could cut a stone ready for the pyramid," he said. "But I'm not ready to explain how just yet."

The room, with its shelves and tables and wooden crates, had a dusky light. Hawass was seated on a wooden carton. "I dislike disturbing mummies," he said, "as that can disturb their afterlife. I feel you shouldn't put them in museums for *thrills,* but for science. I will examine things, document them, and then quickly return them to where we find them. I hope they will forgive us. It is all in the interest of greater human understanding. Knowledge." For a few minutes then, he was silent.

"Can you think of lessons our culture can take from what's been lost in ancient Egypt?" I finally asked. "Are there things that we should write down in case something happens to us? So people in the future know who we are and what we did?"

Hawass sat quietly some more. There was a long pause.

"That requires thinking," he finally said. "How much *could* we write down permanently? And on what do we write it? Stainless steel? Because, you know, paper eventually dissolves. Stone gets eroded. . . ."

He ignited one more of his trademark smiles. "And in what *language* should this information be written? Languages disappear, too. How do we communicate to the future in a way that people might understand? I believe that's one reason the pharaohs had the pyramids built: as a statement. But speaking for us? In today's world? How to communicate to the future? That is something serious to think about."

Since the mid-1980s, no other city in Europe—or maybe the world—has revitalized itself the way Bilbao has.

Two decades ago, the place had been ground down into a shuttered, rusting, postindustrial wasteland. A century prior to that, Bilbao is said to have been Spain's wealthiest city, with one of the nation's largest and busiest ports, which served a vibrant steel, chemical, and shipbuilding industry. Cargo vessels moved in and out of the Bay of Biscay, depositing or picking up goods and materials on Bilbao's shore. But then, in the years after World War II, under Generalissimo Francisco Franco's repressive Republican government, Spain fell subject to trade boycotts and was excluded from both UN and NATO membership, and the local export economy disintegrated. By 1983, when heavy floods decimated much of the old city (or *casco viejo*), city fathers finally decided to reorient the local economy toward tourism and services like communications, higher education, and advertising.

It worked. Today (and with apologies to New York, London, and Hong Kong), Bilbao is the coolest and most energetic city on Earth. Everywhere you look, visual art has been consciously included in the vista. From the eye-catching pinkish red paint covering the city's deconstructionist Sheraton Abandoibarra hotel to outlandishly patterned plots of flowers planted in the city squares to the strange, string-like spans of new bridges crossing the river, the city is forever catching your shocked gaze and watching you marvel. Then it winks back: *Gotcha,* it says.

At the center of Bilbao, with its rounded and uneven architectural stacks of glass walls and cantilevered sharp edges, much of it covered in sinuous, silver fish-scale squares of titanium, is the Guggenheim Museum. Backed by the broad and greenish waters of the river, and owning the landscape around it with broad promenades and squares and parks, there's nothing predictable about the museum's setting. Fronting its entrance is a topiary sculpture that looks like a 43-foot-tall version of a child's stuffed dog. Titled "Puppy" and created by the sculptor Jeff Koons, it is meant to resemble an enormous West Highland white terrier. Made of steel covered with soil and small, blossoming flowers in yellow and red and blue, "Puppy" takes every saccharine symbol imaginable and, all at once, throws them together with an irony that verges on the unholy. And yet, like the reinvention of Bilbao itself, the sculpture completely works.

Each day about lunchtime, having bounced around the city on foot, interacting with Basques in city squares and coffee shops, I find myself gravitating back to the museum to walk past the statue, eventually finding individual tiny flowers in yellow or red or purple that I want to check on. Then I go inside.

One day, I spend an hour in a windowless gallery longer than a football field, where Richard Serra's huge, rusted-steel installation "The Matter of Time" rests on the floor, its arrangement of 12-foot-tall walls like an only slightly comprehensible maze. To experience it, you walk along and between its gigantic tilting steel lines, following them sometimes inside inward-turning spirals, and other times between two undulating walls that seem to almost touch because of your perspective, apparently barring your path ahead, only to actually provide an opening as you grow closer to what appeared to be the point of closure. To walk in and among these shapes is to have your perceptions bent and stretched; your sense of understanding space thrown off. Everything about it, the world around you, the very air itself, is affected by Serra's creation.

All over Bilbao, but especially inside the Guggenheim, I can't help but feel the energizing whiff of endless human creation always there, possible and waiting. On another day, after wandering through some artisan's shops, I stop by the Guggenheim again and spend an hour with the work of the conceptual artist Jenny Holzer. The gallery is three stories tall and strung floor to ceiling with narrow message boards, each running simple sayings and aphorisms in Basque, Spanish, and English from floor to ceiling in red LED-lit letters. "I Say Your Name . . ." they say, the words coursing from out of the floor and toward the sky. "I Save Your Clothes . . . "

Then, every day while still at the museum, I stop by the restaurant on the museum's third floor for what many people say is the best lunch in town. One day I have roasted eggplant, followed by braised tail of veal swaddled in sheets of pappardelle set in a shallow pool of sauce. Another day, it's roasted lamb shoulder and peppers. For dessert, each time I visit, I have a house specialty: inch-tall columns of different flavored ice creams, their exteriors rolled in a layer of ground cocoa and coffee. Each day, before the dessert arrives, the servers clear the table of utensils. No silverware accompanies the ice cream from the kitchen. Everyone has to eat this final course with their hands . . . and before it melts. It's terrific: Dessert becomes a countdown clock.

Again, Bilbao winks back. *Gotcha.*

---

Each day in Bilbao, I spend the early evenings on the Calle de Lopes de Haro, the city's main thoroughfare, near my hotel. There, as soon as the business day's vehicle traffic slows, the broad boulevard is closed for street fairs, complete with jugglers and sellers of food and goods. But there's more. Elevated up on stages that number one or two per city block are living dioramas of famous Basques from history, featuring actual people doing period-clothed *homages*. Finally, eventually, I head slowly down Calle de Lopes de Haro to the river, and back to the Café Boulevard, for a few pintxos and a newspaper.

One evening at the Café Boulevard, as Enrique is looking after me with his usual generosity of spirit, I meet a man named Federico. He is sitting next to me at the bar, and joins in as Enrique and I chatter about events and sports. Federico is probably 55 years old, dressed in a three-piece suit. He's an attorney in town, he says. He, too, is reading a newspaper; his is a local one to my *Herald Tribune,* and there's an item on the open page between us about ETA.

"Are you a Basque?" I ask Federico.

"Oh yes," he says. "Why?"

I tell him of the journey, my history.

"Well, that *is* interesting," Federico says. "We want to welcome you and your DNA back. Bring your whole family. This is a good part of the world."

In the low-light, green-tinted air of the café, Federico asks Enrique for two more glasses of wine. It's his favorite, another red from the Rioja region. "As far as I'm concerned," Federico says, "this wine is the best in the world. It is not expensive, it is light yet warming. It is like the sun on your face." As he describes it, the wine shows up. He makes a toast to my successful journey.

After we toast and sip, I point at the newspaper story between us. "So," I ask. "As a Basque living here, how do you feel about ETA?"

Federico pauses for a long moment, his mouth frowning in distaste. "They're *children,* though many are as old or even older than me," he says. "They are adolescents. Vandals. Terrorists. They don't offer *solutions* to the society's problems they object to, but exist to destroy things and innocent people's lives in the name of their own frustrations. Many, I suspect, are not educated, in the usual sense. They offer nothing other than violence and terror as an option. They only take things away from the rest of us. That is their intention: to take things away from others as retribution for their own frustration and anger."

It's growing dark outside. Federico smokes a small cigar. We sit and talk some more: about sports, the most recent Tour de France, and the loveliness of the Basque countryside. I eat some deep-fried shrimp off a large white crockery plate that Enrique has brought over, and we glance periodically at our newspapers, the conversation moving along when one of us has something to add. And I have to say this: about ETA, Federico has a point.

A few days later, however, I find another perspective.

Having covered many of the appeals of Bilbao, I decide on a field trip to the nearby city of San Sebastián. Built of low skyscrapers and seaside promenades in a crescent of flat land between an elegant and semicircular bay, the Bahía de la Concha, and the steep and verdant hills near where Spain's Guipuzcoa Province meets the French border, San Sebastián is alleged to be the most Basque of all Basque cities. It is also one of the world's premier food towns. So in the name of a little exploration, and the fact that I can report to my "foodie" wife that I've been there, I telephone the restaurant called Arzak, a Michelin three-star cathedral to chow lauded as not only the best restaurant in Basque country, but in all of Spain.

"We have a long list of reservations," a woman says on the telephone. "You should have made plans in advance of today. But cancellations are possible, though they do not happen often. Still, if you are going to be in San Sebastián anyway, contact us at 1 p.m., and we'll see if anything can happen."

I rent a car and drive the hour north by east from Bilbao. Cruising along the highway between the two cities, the Atlantic to my left is roiling and blue, with corduroy lines of whitecaps where the water approaches the beach. Once in San Sebastián, the Bahía de la Concha is knee-bucklingly aquamarine and gorgeous. And within an hour of parking the car, I'm rolling up my trouser legs and wading knee-deep into the bay's blue-green shimmer, the city's stately line of white-faced hotels and office buildings arrayed out behind me.

San Sebastián and its *bahía* play a part in Ernest Hemingway's novel *The Sun Also Rises*. It's the place where Paris-based expats from the "Lost Generation" end up, having spent the week before that debauching themselves at the Fiesta de San Fermin in the town of Pamplona, up in the hills above San

Sebastián. At Pamplona, Hemingway's characters, all of whom have been traumatized in various ways by the shocking brutality of World War I and the unleashing of the Modern World, distract themselves by running ahead of the bulls each day, leading them from their pens at the edge of town into the bullfighting stadium. Then Hemingway's characters stay to watch the bullfights, eat and drink far more than they should, and eventually destroy their relationships. It's a book entirely about the horrors of World War I, where the war is rarely mentioned. And despite a not-so-pretty story line, it remains a literary masterpiece and a sensation among a certain youthful and largely male demographic today, nine decades after it was published.

On this specific day, in fact, evidence of the book's enduring influence is scattered everywhere along San Sebastián's bayside promenade. The fiesta, it turns out, ended last week, and scores of young men dressed in the festival's traditional white trousers, striped Basque shirts or white button-downs, and red bandana neckerchiefs have finally come down from the mountains and lie sprawled, passed out or hungover, on the promenade's sand and grasses. They're asleep on a majority of the promenade's long and seaward-facing benches, as well. Many are covered head to toe in stained splotches of red wine. They look like human wreckage. It's as if a giant has come through and hurled human bodies everywhere; which, I guess, he has.

There was probably a time when such an experience would have appealed to me, but, for better or worse, that time has passed. Still, they've all gotten a memory, and probably a story or two, from their time in Hemingway Land. Now they can return to London or Palo Alto, one more rite scratched into the notebook.

Unfortunately for me, when I contact Arzak at 1 p.m., the dining slate is full. "Even the small table in the kitchen is being used today," the hostess says. "This is where I hoped you could eat."

On the menu are things such as lobster with white olive oil and toasted lotus root with fish mousse. I cannot even begin to know what those taste like. Bits of what look like lamb loin rest in pools of sauce in circular indents of white plates; the food sits so perfectly it appears a Broadway set designer must work in Arzak's kitchen. Supposedly, back in the kitchen, the chefs use ingredients like liquid nitrogen and vacuum-sealing bags to render their dishes unique in the world. The place itself, while humble on the street level outside, is so clean-edged and industrially, spotlessly slick inside—complete with lighting that seems to cast no shadows—that computer chips could probably be manufactured in there. And yet, the people working at Arzak seem genuinely nice, and even sad that there's no extra seating.

So, instead, I find a good pintxo place near the old cathedral in the oldest part of the city not far from the waterfront, and settle in. Smoked hams hang from the ceiling. There are anchovy pintxos, and more baby eels with pimento. There is Manchego cheese and *jamon.* There are these green olives strung three apiece on wooden toothpicks, each olive filled with cream cheese and rippingly spicy pimento. An eater could do far worse.

Then, after an hour or so of poking around in town, where I buy an antique running block from a sailing ship's rigging as a souvenir, it's time to head back home. On the way, in the interest of catching a little more of Basque Spain's vibe, I want to drive into the hills, to see what Enrique is talking about when he describes the sun and the green roll of the landscape.

Once out of town, the roads up into the hills peel off into smaller and smaller tracks. Soon I'm somewhere between San Sebastián and Bilbao, the ocean a long way in the distance, with its sun-shot blues peeking out when there are valleys between mountainsides. There are vineyards and spreading pastures. There are small plots for crops. The day is pleasantly warm, so I have the car windows open; the radio is on a pop music station.

Then, cresting a hill, there's a small, ancient-looking church, with two stone buildings next to it. The road ends there, so I park, get out, and look around. It's gorgeous. The church is abandoned, the roof on its rocked bell tower appears to be slowly rotting but still solid, and the old buildings appear inhabited. A radio is playing music somewhere. There is a wooden table with a large, market-style umbrella poking through its center. There are folding chairs. The green grasses across this place are almost knee-high, and even in the afternoon the tall blades of grass still hold enough dew that my trousers are dampened.

After a few minutes of looking around, I see a man watching me from the second-story porch of the far house. He's dressed in blue jeans and a thick, black wool sweater that appears to have been knitted by hand. He's perhaps 30, and his hair is cut eccentrically, with patches taken from it.

"Hello," I say.

He waves. "Hello."

I walk over to where he is. As I do, he comes down the exterior stairway of laid stone from the second floor.

"I'm just looking around, exploring. From America."

"Welcome," he says.

As usual, I explain what the journey is about.

"That's cool," he says. "I'm a Basque. I grew up here, in the Basque country. Though not truly nearby to here."

"So do you live here now?"

"Yes."

"Why did you choose this place?"

"Because these buildings are abandoned. It is free."

"How long have you been here?"

"About six months. I live here with seven other friends."

Looking around, I see that no wires lead to these buildings. They're basically piles of rock, their roofs complete but probably leaky in the rain or snow. Chimneys poke through. There are piles of gnarled firewood and a few split logs tucked against the houses, protected by the eaves; a small vegetable garden grows between the two buildings.

"Do you have electricity?" I ask.

"Yes," he says. "Solar and batteries. But none from the outside."

"What about water?"

"There is a well nearby. We carry it. We have a good propane stove. It's a good place. It's beautiful here. Peaceful."

I look around. He's not kidding. In every direction, beneath the sunshine, the mountains and hills roll away, green and rich-looking. Far in the distance, the huge blue ocean glints. "Yes, it is beautiful," I say.

We introduce ourselves. His name is Miguel. And over the next 20 or 30 minutes as we talk, it becomes pretty obvious that if he's not a Basque nationalist and full-fledged member of ETA, then he certainly leans in their direction. And yet, he seems reasonable. "You know," Miguel is saying, "For a time my friends and I wanted to reject the world that was forced upon us, the life and expectations forced upon us. The idea that the people in power say I *must* do this . . . I *must* do that. I want my freedom. Why should I live in a world with electric bills? Why must I drink only milk approved as safe by the EU [European Union]? What's wrong with the milk from my neighbor's cow, as my ancestors drank it?"

"But what about these houses? You didn't build them? You're benefiting from those who came before you?"

"No, these are quite old. But since no one is using them, why can we not live here? We are left alone very much of the time. You are the first stranger to come here in weeks, actually."

"What do you do for food?"

"We all work jobs. We live communally. Among us, we have two cars and a motorcycle. All of them are away today, which is why I am here by myself. If one of the cars or the motorcycle was here, I'd drive to the village to do some work and buy some small supplies for the next few days."

Screwing up my courage, I finally ask. "So, are you a Basque separatist? A member of ETA?"

Miguel smiles; the dark, day-old stubble on his cheeks crumples into dimpled folds, his dark eyes go into a squint. "Well," he says, he chuckles and shrugs, "I support what they are trying to *accomplish* . . ."

"And what is that, exactly?"

"Freedom."

"Freedom from what?"

"Freedom from others telling us what to do . . . to think . . . how to feel."

"And you think the government here and in France can do that, dictate to you what your life will be?"

"I know they can."

"How?"

"Okay, an example. How is it that drug companies decide what my emotions should be, and the government then allows these companies to sell the drugs that will make me feel this way? Why can I not feel angry or happy or depressed when this is how I feel? They want to moderate emotions through drugs. It is not natural."

"And there are other things like that?"

"Many. Everywhere you look. Mandatory conscription into the Army? Certain kinds of taxes? And if you don't agree, they don't want to talk about that disagreement, they put you in jail . . . or throw you into the Army."

"It's more complicated than that."

"Yes, it is."

"What about the road you use to go to town? Taxes paid for that."

"Do you really think there weren't roads before there were *taxes*?" Miguel smiles. "But the point is always the same. If you disagree with power, you are pushed under their authority. For government, the use of threat or violence is just another form of politics."

"So violence is justified for the other side also?"

Miguel shrugs. "Well, no," he says. "But governments use pressure and violence all the time to enforce their views, so . . ." he shrugs again. "Once in a while, if the damage can be contained, maybe actions get a point across? At the G7 or G20 economic conferences, these governments and their leaders don't want protesters, so they erect fences. If the protesters get past the fences, the police try to drive them back with water cannons. If the water cannons don't work, they hit them with tear gas and clubs. We have seen this to be true. All because the protesters want to make a point that the governments don't want to hear."

There's something more than a little crazy about all this. I am standing on this beautiful hilltop, surrounded by buildings erected 500 or 600 years ago, one of which is an ancient church, talking about the value of terrorist violence with a guy who, inexplicably, makes some sense.

How can this be?

"The point is," Miguel continues, "we have come to live in a world where anyone born new into it has had many personal choices already removed. The only question is, depending on that person's sex and where that person is born, which free choices remain? Taxes. Religion. Economic status due to a global marketplace? As a baby, you arrive in the world inside a box that was built for you before you existed. So we say: We are people, and we want our freedom back."

We stand in the courtyard for a few minutes more. Taking in the place, chatting. I can almost feel the afternoon sun heating the stones of the buildings, which will help keep them warm tonight against the altitude and the overnight mist of coastal Spain. This place is both stunningly beautiful and tranquil. Green hills stretch away in every direction. I can't get enough of just looking at it. The rough stone buildings. The blue ocean on the horizon to the west beneath a sky of different blue. The clean yellow of the sun. It's like being inside a kindergartner's crayon drawing.

"If you would like," Miguel says. "I have some coffee in the house that I made an hour or two ago. We could reheat it and take a cup."

"No," I say. "It's okay. For me, it's time to go back to Bilbao. But thank you. "

"Of course," he says.

As I walk back to my rental car through the tall grass, Miguel follows, stopping halfway between the house and the car. He waves. "Go safely," he says.

The next morning at 7 a.m. I open the French doors in my hotel room, which overlooks the city's central Plaza de Federico Moyúa. The view is spectacular. From the opened doors, I look down on the giant traffic circle in the sunrise; at its center is a city park built around a circular fountain. The plaza is its own work of art, with its outer areas transected by straight lines of sidewalks, and deep-green lawns, and optically complex plantings of flowers: blooms of white and red and black that seem to create jagged wave patterns to a viewer with a little elevation. Though the new day is still early by American standards, Bilbao's people are everywhere: crossing the plaza, waiting for a bus or hailing a taxi, going down into the subway entrance across the way.

All of these people are dressed and ready for a morning of work. Maybe it's their ancient farming heritage, or maybe it's only that they live in such a beautiful place and plan to enjoy their surroundings later in the day, but the people of Bilbao start their workday early; then they end it the same way.

Each afternoon at about two o'clock, all across the Basque region, natives break for lunch, their largest meal of the day. Then, unless events are truly pressing at the office, people's cares are left behind, and it's off for a siesta, or to do some fishing or shopping, or perhaps some cycling along ancient networks of roads that crisscross the region's hillsides. Trust me when I tell you: On a sunny weekday afternoon outside Bilbao, the roads are thick with people in yellow and black and red spandex, all riding really nice road bikes.

After phoning downstairs for a cup of room service coffee, I go back to the open French doors to stand and stare. Without sounding too much like the "take away" message at the end of too many episodes in the original *Star Trek* TV series, modern humans, across the last 7,500 years or so, have become capable of amazing and terrible things.

We have cured diseases and waged wars, flown to other planets and plumbed the depths of the seas, and we can now send waterskiing photos around the world in the blink of an eye.

We have thrived and procreated until, according to many people, we're taxing the planet beyond what it can tolerate. And yet, even inside this huge, diverse, and sometimes miraculous *Homo sapiens* experiment, which has so far lasted a fraction of the time of the Age of Dinosaurs, the tug of war inside each of us never stops.

"I'd say that a human population of about 200 million on Earth is in the right ballpark," the neurophysiologist Dr. William Calvin said. "It could be more, it could be some less, but 200 million on Earth sounds about right if we are to live in balance with the environment."

Though Calvin is a professor at the University of Washington's School of Medicine in Seattle, and over the last 20 years he has published several best-selling books about the evolution of human cognition and how brains think, he has also become a national authority on the Earth's environment and the human toll on it. When I caught up with him, to talk about the long-time rise of humans and their penchant for finding new technology traps, Calvin seemed more interested in talking about the drag human technology was placing on the environment.

"The vast amount of $CO_2$ our cars and industries and processes have pumped into the atmosphere, well, there's nowhere for it to go. All the historical models show that we should be headed slowly into a new, though not terribly steep, ice age. Instead, because of all the carbon in the atmosphere due to a fossil-fuel economy that's only rising around the world, that ice age has been postponed. Maybe indefinitely." To Calvin, what's happening to the environment and the technology traps that human societies have found themselves in for millennia now all stem from the same, deeply basic human cognitive trait.

"It's a very simple human thought pathway, and one that I expect engaged itself very early in our development," he said. "You learn how to do something, you get better at doing it, and then you try to improve it. It's a simple question of repeated behavior and repeated reward. It's an incremental path. You don't think far ahead, you're just doing the next thing. But then, eventually, you start getting diminishing returns. The people of Mesopotamia didn't realize until too late that salt had left their soil degraded. One generation to the next probably couldn't even see it; didn't figure it out. These sorts of changes, they happen very slowly."

Calvin paused. "You know, as a species we're intelligent and resourceful, but it's hard for anyone to really see ahead a great distance in time. Since the rise of farming and early language, we've always had teachings and schools of some sort. And in these schools, we try to educate children, instructing them to look ahead. We say: *Think* about the outcomes, the consequences. But who can really see much farther ahead than today and tomorrow's cause and effect? It's basic human cognitive behavior, as it's been since the beginning. If things are working, we keep doing them. Which is where the problems sometimes start. Can we realistically see ahead a century? Five centuries? Ten? At some point, I mean, come on . . ."

Still, this kind of blithely blind allegiance to one approach, one *way,* is what has led several societies before ours into destructive technology traps. And as Calvin wants people to know, even though modern humans, with global trade and hop-scotching areas of famine and drought, are not ancient Mesopotamians or the Maya, our modern cultures and technologies may be no better at avoiding future technology traps than were some of our ancestors.

"Keeping things diverse, in terms of what a society is doing, that helps," he says. "So diversity is something to look for. But if some technology or line of thinking is working, it's human nature to keep using it, with more and more people adapting to it at the same time. We're thinkers and watchers. We recognize the relationship between some form of action and its reward. And even with all we're soon going to have to deal with environmentally, I'm feeling somewhat optimistic for our future. I think a lot about human creativity. There's reason to be hopeful because of that."

The Neolithic revolution's population explosion, Calvin believes, happened because humans had grown able to express and distribute their thoughts. "We'd finally gotten enough experience that we could grow ideas. We looked at the environment around us. We watched. We thought about it. We had people generating Plan Bs. That requires elaborate thinking. And, really, ever since, all we've been doing is making our Plan Bs more complex as society grows more complex, as well."

Not that all Plan Bs are good. In the 1890s, for example, Alfred B. Nobel, inventor of blasting caps, dynamite, and smokeless gunpowder, believed the arms race between nations his inventions had helped set into motion would become human salvation. In a kind of lockstep, many of the world's advanced nations believed that if they built enough weapons, no one would dare to wage war. In arming themselves and selling these weapons to others seeking protection, many of these nations grew rich alongside Nobel.

"My factories may make an end of war sooner than your Congresses," Nobel wrote to the peace crusader Bertha von Suttner in 1892, "because the day two armies have the capacity to annihilate each other within a few seconds, it is likely that all civilized nations will turn their backs on warfare."

The Plan B in that case became World War I.

"So, as I say," Calvin says, "while my training is in human neurological pathways and why people have come to think the way they do, I've somehow ended up looking at the effects of people on the Earth. As things stand now, the planet can't keep sustaining this many human beings for an extended period of time. We're already seeing the crushing need for resources in some places creating conflicts, even wars, between the haves and have-nots. And because of the resources and technologies associated, I think the haves are going to win out. In the end, if you ask me, humans and the planet will go on, but with fewer people. And as I said before, 200 million is about the right number, down from the six and a half billion today. Of course, it won't happen all at once, and getting there is fraught with wrenching tragedy. Some of it will be environmental. Some not. But what we have now, it's ecologically unsustainable. The amount of $CO_2$ we've pumped into the atmosphere. A petroleum-based world economy facing diminishing amounts of petroleum? The collision is coming. In some places, the collision is already here. Just read a newspaper."

When I tell Calvin of my flight from Washington Dulles to Frankfurt, complete with wine brought over from France only to be flown back to Europe and Gala apples brought in by refrigerated vehicles, he gives a deep-throated "heh-*heh*" chuckle. "So you see the problem?" he said. "A petroleum-based economy, with more and more people and cultures competing for that same diminishing resource? And all this inefficiency already in the system? How long can *that* keep going? How is *that* sustainable over any real length of time?"

The slightly scary part of listening to Dr. Calvin is that, unknown to him, many scientists, assisted by UN reports, have also generated an idea of what a "sustainable human population" on Earth is. And that figure also happens to be roughly 200 million people: a population lodged into traditional hunting-gathering lands or distributed across some of the most habitable spots. This figure, incidentally, is only 3.5 percent of the world's current head count.

Ironically, in late June 2008, I'd spent several days visiting a place where roughly 3.5 percent of the population remained: the Chernobyl Exclusion Zone. There, in late April 1986, during a routine safety check, a nuclear reaction ran out of control at the Chernobyl power plant in the Ukraine, with the reactor literally blowing its 640-ton top due to overwhelming pressures inside. When air hit the nuclear core there was a second explosion, throwing a sparking, spitting fireball a mile into the sky and hurling fragments of burning nuclear fuel across the landscape, ushering in an atomic nightmare. At least nine tons of radioactive uranium were spewed into the sky, 200 times the amount released at Hiroshima.

Two days after the explosion, Chernobyl's radioactive smoke had drifted 700 miles northwest, far enough to be detected by workers at a Swedish nuclear plant, which forced the Soviets into a public declaration. With their announcement, they also ordered the evacuation of dozens of towns near the reactor, displacing 125,000 people. Over the coming weeks, 93 towns and 175,000 people would be displaced, never to return. In the end, two people were killed in the explosion, another 29 firefighters died in the next few weeks from radiation exposure, and 6,400 casualties were attributed to the accident in the form of cancer and leukemia. Eventually, all Chernobyl reactors were closed. To isolate the most radioactive areas from human contact, the Soviet government drew two concentric circles around the bull's-eye of the fire zone. The first, called the Inner Zone, has a 10-kilometer (or 6.2-mile) radius. The second, with a 30-kilometer (18.6-mile) radius from the reactor, is called the Exclusion Zone—or, with a nice Soviet-style flourish, the Zone of Alienation.

In June 2008, 22 years after the explosion, the Zone of Alienation was still in place, its 21,000 square miles still inside a sentry-guarded cordon, its former housing for 175,000 people in 93 towns and villages off-limits forever. If you're lucky enough to visit the Zone of Exclusion, you need both an official guide and a "minder," who lives inside the zone alongside roughly 3,400 others charged with keeping the stilled reactors maintained. In addition, about 300 illegal squatters live there, elderly people who moved back in defiance of the evacuation orders. Visitors also must sign a contract explicitly stating that they "agree that the State Department–Administration of the Exclusion Zone shall not be liable for possible further deterioration of their health as a result of the visit to the Exclusion Zone."

This was not comforting. Still, to visit a world whose population has evaporated was both freakishly interesting and sobering. Visiting abandoned collective farming villages of Cherevach (former pop. 460) or Zalissya (former pop. 2,849), stepping past the yellow triangular signs with red "radioactive" symbols that line the roads, what was most impressive were not the trees growing up through the interiors of the little dacha houses, lifting their roofs off and leaving each house disintegrating as it twists off its foundation. Instead, it was that the Earth's force of life remained so visibly strong. Wild animals had returned everywhere. Short-toed eagles with heavy gold wings, elk with massive racks of antlers, red foxes, lynx, and *lots* of wild boar. All of them, plus hundreds more species, had moved back into places once occupied by people, including towns and cities. Raw nature was reacquiring what man had owned and civilized, and nature was doing it through a rusting, rotting, relentless grasping back. It was more like watching a slow flood than a stunning blast.

Among the strangest stops inside the Exclusion Zone was Prypiat. There, a city of stolid 15-story apartment blocks that was once home to the 48,000 engineers and custodians central to maintaining the Chernobyl reactors now sits empty. To wander inside its city squares or abandoned amusement park, with its 150-foot ferris wheel, bumper-car arena, and motorized rides overgrown with tangled trees and weeds, was to be reminded of the loneliness of a world without people. One noon, as I enjoyed a picnic lunch in a city square, the wind began to pick up. The only noise beyond the gusts that buffeted the leafy trees was the slamming of unsecured doors as drafts slipped in through open balcony doors or shattered windows and raced along the empty apartment building hallways.

In the slamming of those doors, I heard the sound of a reduction of human population by 96.5 percent.

To put a human face on what happened at Chernobyl in 1986, late one afternoon in the Exclusion Zone, my handlers and I drove out to Paryshir, an abandoned village inside the Exclusion Zone near the Ukrainian northern border with Belarus. As our SUV streaked along another empty and slowly breaking-down stretch of pavement, a brownish wild boar—its body dense-looking and round, its head seeming too small for its body, as all hog heads do—stood at the side of the road, watching as we passed. "That's a young one," said Vladimir Verbitskiy, an officer at Chernobyl's regional control center and my local minder inside the zone. "The mature ones are black. They're everywhere."

As we passed above the Prypiat River on a deserted, poured-concrete bridge, a large painted turtle crossed the empty road. A green floodplain stretched in all directions. We were headed to meet one of the 300 illegal resettlers. My minders told me that, usually, officials don't bother the squatters. "Many are not friendly," said Vladimir. "Still, though it's illegal to resettle, the authorities have finally said okay."

The SUV slowed and turned right onto a side road. The road was lined with little houses, many overgrown with brush or overtaken by trees. There was one, however, freshly painted white, with green shutters, and well tended. It's where biologists stay when they visit to make reports. Our car followed a dirt two-track into untended fields, bumping past patches of trees. Just ahead, at a board-fenced dacha, an old woman wearing a red dress, blue apron, and bright yellow head scarf was sitting peacefully on a bench in the shade of a tree. Just inside the fence, chickens wandered the property. A large, tended vegetable garden sat behind the house, its furrows straight.

"This is Maria Shylan, 78 years old," my guide said. As he shut off the vehicle, Maria stood.

"Come . . . come," she said, clearly happy and excited to have visitors. She opened the wooden gate and led us into her yard. "Let me show you my garden . . . though I'm fighting the potato bugs at the moment. I'm the only one here to fight them. In the end, the bugs may get the potatoes and the wolves will get me."

She led us back to the garden, which covered about an acre. "I have tomatoes, potatoes, onions, carrots, green vegetables, anything you might need. I eat some, and preserve some for the winter," Maria said. "Scientists come to check my crops, to make sure they're safe to eat. They check the water in my well, too, to make sure the drinking water is safe. It is. I have pigs and chickens for meat and eggs," she said, gesturing toward a muddy pen and a full chicken coop. "And about once a month, a salesman in a grocery truck comes, and I buy the few things I need: sugar mostly. Once a week, a bakery truck comes, too, but the bread gets hard in two days, so once a week isn't enough. Mostly, my bread is potato pancakes that I make myself."

Is she here alone?

"I used to have a little dog. But the wolves got him. You see wolves all the time, they come out at dawn and dusk. One night my little dog was barking, so I knew something was around. Then it grew quiet, and my dog was gone. I never saw him again."

Standing in the sun, fists on her hips, Maria was smiling a near-toothless grin. "Sorry about my teeth," she said, "but they all fell out after the explosion, which we couldn't see from here. But, the next day, there was much activity. They loaded us all onto buses, and moved us away, to Kiev. Many people went into older-care homes. Not me. After a time, I grew homesick, so I came back. This house is where my family has always lived: my father and his father. Now I have all this to myself. I enjoy this simple life, though it is lonely."

Suddenly, tears came to her eyes. "Come," she said. "I have something for you. "

Maria headed into the house, her walk an energetic waddle. The house was spotlessly clean. The ceilings were low, and the plaster walls were covered in bright blue paint. Curtains partitioned off different areas, likely to better retain the huge earthen stove's heat in winter. Maria motioned for us to sit at the kitchen table. Then, from a wooden cabinet, she pulled out a plate of potato pancakes ("I made these just today"), a big bowl of honey, and a jar with some clear liquid inside, plus two small glasses.

"Eat," she said. She grabbed my hand, placed a pressed metal fork into it, and stabbed a potato cake with the fork, then dunked it into the honey. "Eat. I love to feed guests."

Then she poured two small glasses of liquid and motioned for us to lift and drink. Maria and I both toasted to her health simultaneously. The liquid smelled like water and burned like Satan. It was moonshine. Vodka. My eyes clouded up, my stomach recoiled. To settle my belly, I took a bite of the potato pancake and honey. It was dense but tasty. Sort of like corn bread.

Maria refilled the glasses. "Again," she said. "I make this myself. I put sugar, potatoes, and bread with yeast and water in a jar. It takes two weeks of brewing, then I cook it and strain it. In summer, I work outside to make this place better. In winter, I have the television. It's easy enough, though I don't have my health."

Having spent the day gardening, then knocking back shots of moonshine, Maria appeared healthier than many 40-year-olds I know. But the day was growing late; the sun was now slanting into the house at a low angle, and Maria was winding down from her day's efforts. "Stay for the night if you'd like," she said. "To be alone in a place like this is to be *very* alone."

I told her that we had to go. (Truth is, we were staying in a new, radiation-shielded hotel in Chernobyl, and by law we had to get back to the hotel and sign in before dark.)

As she walked me and my minders out to the car, Maria stopped, looked across her gardens and chickens and pigpen, and chuckled. "I like you," she said. "Come back and visit me again soon. Before the wolves get me."

As we drove back to town and the hotel, all I could think was: While I always believe I'd like to live out in the wilderness by myself, growing vegetables and hunting for my own food and living in balance with nature—with wolves at the edges of my life—Maria showed me such a life is more appealing as an idea than as an isolated and lonely reality.

In fact, my visit to Chernobyl showed how much I like grocery stores and wireless Internet and day-to-day human interactions.

Standing in the French doors of my room in the Carlton Hotel, overlooking the Plaza de Federico Moyúa in Bilbao's new morning light, I can't find the stomach to roll the tape forward, or back, to a world that contains only 3.5 percent of the current human population. After all, if the rise of *Homo sapiens* has taught us anything, it's that we are resourceful beyond imagination. And that's true now more than ever. After all, more humans implies more resourcefulness, more imagination.

Humans named the electron in 1894, believing it had no practical use. Today, free electrons power our houses and are central to the hard drive on the computer I'm using to write this book. Where clay tablets were once the source of record-keeping and money, replaced by coins and later printed cash, much of commerce is now transacted through the online transfer of electrons. To anyone using a credit or debit card, electrons have replaced the idea of hard currency almost completely: just another step further out onto the scaffold of economic exchange.

And yet the human embrace of the electron and its potential is only one aspect of the ways, right now, people continue to change and grow their control over the Earth. Right alongside the rising "hive mind" of the Internet are the dramatic breakthroughs in genetics and science, the further exploration of space and the oceans, and the powerful physical, computational, and digital imaging tools that make the world more understandable, tools that didn't exist just a few years ago. Knowledge and the ability to employ it is logarithmic, building on itself with each new layer of what's understood and how it can be used. Over time, with successive layers of understanding, the pace at which we discover what's knowable and achievable quickens. Today, we are learning at a speed never before experienced.

To think about this in another way, there are far more computer processors in the average American automobile in 2010 than were aboard Apollo 11's lunar excursion module in 1969, the first human-piloted spacecraft to land on the moon.

In the 72,000 years since the Mount Toba catastrophe, modern humans have grown and learned and prospered in fits and starts, our discoveries working themselves along equally balanced positive and negative paths. Always, we keep finding our way, experimenting. And inside each of us is virtually the same DNA that left our ancestors thrilled and amazed to have created a stone tool . . . and that today allows us to research new vaccines or fly across the oceans at 36,000 feet, comfortably enjoying life in the frigid and airless upper atmosphere, an iPod in our pocket, a glass of good French wine in hand, and a touch screen video console ready to entertain us for the ride.

As a species, modern humans have now walked far out onto the scaffold of ideas. And while it's become fashionable to talk of societal collapse and global warming and overpopulation, of pathogens and terrorism, if human history has proved anything, it's that trial and error, plus a little desperate resourcefulness, can bring us through the tight squeezes.

And what if the furies are headed our way with nothing but bad news? Well, then, if we follow the path we've taken over the last roughly 6,000 years, the human experiment in specialized societies and multilateral trade between relative strangers amid increasing technologies will likely adjust.

As my visit to the Chernobyl Exclusion Zone showed, the greater, non-species-specific life of the Earth will likely go on.

Personally, and taking into account Dr. Calvin's very valid point that no one can predict outcomes too far into the future, if I had to bet on a population most likely to control the world for thousands of years to come, I'd still put my money on *Homo sapiens.*

Standing in the French doors and staring out across the busy morning as the people of Bilbao head off to work beneath the new day's sun, I wait for my coffee and find myself thinking of Mohamed Wehbe, one of the guides at the ruins in Baalbek. At the moment, something he said seems truer than ever: "Each morning," he reminded me, "is something like a gift."

Might the future get harder for us and our children? Yes. May parts of what awaits be terribly difficult? Possibly. Is it worth it? Absolutely.

We can do this.

Or as William Shakespeare—a guy who understood a few things about fate, history, and human nature—once put it, roughly quoted, our future lies not in our stars, but in our hearts.

On my last day in Bilbao, I head to the Guggenheim for one final visit, one last pass by Jeff Koons's 40-foot-tall puppy sculpture, coated in flowers, outside the museum's entrance.

Inside, the vast galleries and open spaces spread away in every direction, over several levels. Exhausted and wandering, I end up in a room I haven't visited before. There, taking up much of the far wall, hangs an enormous painting that seems to drag me toward it. Called "The Land of the Two Rivers," by the German artist Anselm Kiefer, the image is 23 feet wide and 12 feet tall. Epic and intended to represent the valley between the Tigris and Euphrates, the work is made from electrolyzed salt on copper and zinc, and it has a strange, entrancing, shimmering quality. According to a description of the painting, "This monumental work features an inscription with the names of the rivers Tigris and Euphrates, a reference to the Mesopotamian civilization in whose banks settled some of the most ancient peoples in humankind. Records have gone beyond the decay and ruins of these cultures and have made their past a present . . ."

Their past a *present,* with the salt in the paint again echoing the problem that, in the end, undermined the Mesopotamians. In shades of gray and brown and black and pale white: The earth of Kiefer's vision of Mesopotamia has also "gone white." After this long journey, I feel compelled to stand there, moving first closer then farther away from this depiction of the destruction of an ancient culture, trying to see how it was made.

Finally, after 15 or 20 minutes, satisfied that I have taken this in, I turn away. In another area of the gallery, under a huge bell jar in the middle of the floor, I see what *has to be* Julius's hut, transported here from Tanzania.

Called "Unreal City, Nineteen Hundred Eighty-Nine," by the sculptor Mario Merz, it is an installation that seems—across one, round, 60-foot section of floor—to conjure my entire trip. Covered by a dome-like enclosure of glass panes, it encompasses successively smaller domes of gleaming plastic and metal, as well as, near its front, an armature of metal covered with, you guessed it, the brown thatch of Africa.

Once again, just as at the Registan, my jaw drops.

The installation looks so much like a Hadzabe hut that I expect Julius or another of the clan members to step from it at any second, my Petzl headlamp crimped around his forehead as he munches a chunk of steaming, bright yellow, and recently cooked tuber.

My brain moves a second too slow. It can't reach down and stop a burst of surprised laughter that explodes from my throat: *Ha!* What had been a rudimentary human shelter in Tanzania almost six weeks ago is now replicated inside in a multimillion-dollar art gallery in Spain, right across from a depiction of the despoiled Mesopotamian countryside.

Maybe I'm tired. Maybe it's that, here, at the end of my trip, the scene in front of me has blindsided me with another of Bilbao's supersonic *Gotchas*. But as I stand in the cool, pale light of the Guggenheim Museum, retracing this odyssey through DNA and human history, a jumble of phrases is crashing through my mind. And it's leaving me grinning like an idiot:

We are all descendants of this tree . . .

Life is short, but the work goes on forever . . .

We are running modern software on 100,000-year-old computers . . .

The future starts every day . . .

So here we all are, stuck on this airplane, together . . .

Look at us: *We are a circus!*

We are a circus, indeed; one now wandering the no-man's-land between the desire for safe and established social order and a bloody-from-fingertips-to-elbows howl of "We want our *freedom!*" And so, day after day, the battle goes on inside each of us, the warring sides as old as our genetic history itself.

I've met the family, looked my ancestors in the eye, and, in the end I've come to be reassured by what I saw. Standing all alone in that museum gallery, with history and art and memories of a world full of new friendships and experiences, I am smiling.

Two hundred *million* people in the future, down from 6.5 *billion*?

I think not.

*Home once again, my beloved Blue Ridge Mountains in view*

# HOME

**IT'S A PEACH-COLORED** summer sunset in the Piedmont of central Virginia. As I arrive home, the rolling hillsides and Blue Ridge Mountains to the west also feel like long-lost family members. In meeting the family, I've been almost six weeks away. As I ride in a taxi from the local airport to my house, the car's worn and squeaky fan belt shrieks beneath the hood, disturbing the early evening silence and reminding me of the entropy that hides everywhere in the world, always ready to step from behind a tree to whack you with the cost of existence.

Still, it's wonderful to be home. After flying for almost two days from Bilbao to Frankfurt to Washington Dulles to Charlottesville, in a state of slightly exhausted mental peace and ease, the automobile sound track of this last leg home—about 20 miles of roads across fields and small woods—carries along with it the fan belt's incessant *wheeer . . . wheeer . . . wheer.*

"Gotta get that belt fixed tomorrow," the driver says.

"Or maybe it'll just wear down a bit more and settle into a new pattern," I say back.

"Nah," the driver says. "You should never hide from problems."

The driver is right.

Across the length of this trip home, about 8,000 miles, I've been thinking pretty much nonstop about everything I've seen and experienced in the last six weeks. As I boarded the airplane in Spain, some of it was making sense, but other bits sat outside any form of comprehension, like unread stretches of DNA or shattered bits of once useful crockery.

Still, one fact remained. According to Spencer Wells and others, between the new replication-error genetic markers being regularly added to all of our DNA over time—introducing minuscule new differences—our overall genetic code hasn't changed in something like 100,000 years. We all carry much the same DNA that our ancestors first transported out of east-central Africa's Rift Valley some 70,000 years ago. Yes, some of us are taller, some of us are paler-skinned (so our bodies can absorb more vitamin D from the sun in higher latitudes), but our DNA remains very much the same.

So how can we explain the rise from small tribes of hunter-gatherers in the Rift Valley to people on airplanes flying around the world in the course of a single day?

In the end, the answer has been sitting in front of me all along.

We modern humans haven't so much evolved as adapted. In fact, if *Homo sapiens* are anything, they're adaptive as opposed

to evolutionary. Over the course of our history, personal and environmental stresses have forced us to do things differently: to move, to fight, to seek new forms of food. These decisions have led us into gardens of forking paths, and, often as not, have kept us alive. Sometimes the right path is chosen. Sometimes disaster awaits. But, at its very basis, the bedrock ability to, in William Calvin's words, *watch* and *think* has carried us much farther than have the basic genetics that created us.

*Wheer . . . wheer . . . wheer,* the engine keeps screaming. The driver knows where he is going thanks to a GPS mounted on his dashboard: a tiny digital map and locating computer, itself in a nonstop conversation with, at minimum, five satellites in the sky, just a few of the hundreds of imitation stars we humans have placed up there, 12,000 miles above the Earth, to help us find our way.

"Gotta get that belt fixed first thing tomorrow," the driver says.

And then the taxi is pulling down my small road, passing the neighbors, pulling past my sprawling juniper hedges and into the circular driveway in front of my house.

"You're home," the driver says.

"Yep. Thanks."

After I pay him and we've pulled my bags from the vehicle's trunk and set them on the driveway's asphalt, the taxi pulls away, leaving me to stare at the 70 or so feet of flagstone walkway between myself and home.

It's something after 7 p.m. The house is quiet. My wife and children will be back in a few days, toting the dogs in the station wagon's rear cargo area amid battered nylon duffel bags and piles of dirty laundry. Until then, I have time to think and rest and go through my thoughts and scribbled books of notes, wondering about all I've seen and setting aside the Three Big Questions again for a while as I scratch the cats—who don't travel—behind their ears.

As I slip the leather strap of my briefcase over my left shoulder, then crouch to lift both duffel bags from the ground by their looping handles for the last few steps home, I am smiling. Because the final thing I've come to understand on this trip to my extended family is this: Any good journey not only takes you places you never expected to go, it also leads you all the way back to your own front door.

# ACKNOWLEDGMENTS AND NOTES

**FIRST, I NEED TO THANK** my children, James and Anna, whose births in 1993 and 1994 got me thinking seriously about the intersection of past and present that exists around us every day, should we scratch the surface to look at it. For that gift of insight, not to mention your daily presence in the lives of your mother and me, this book is dedicated to you. You guys are the next steps along a track that stretches back to Africa's Rift Valley, more than 2.4 million years ago.

Also to my wife Janet, who is patient and kind and supportive—not to mention fun!—in ways that leave me grateful every day. She is the miracle that found me.

I need also to thank my parents, Jim and Joan Webster, in New Mexico and Michigan, and Janet's folks, Roger and Jean Chisholm, in Arkansas, both sets of whom are always there for us.

A huge thank-you also goes out to Spencer Wells, Ph.D., and everyone at the Genographic Project at National Geographic. Without Spencer's help—not to mention his input into my manuscript and the book's foreword—the object you now hold in your hands wouldn't be possible.

There's also a world of magazine editors to thank for keeping assignments flowing during the writing of this book, work that allowed the household ship to remain afloat. Chief among these are Keith Bellows, Paul Martin, Scott Stuckey, and Jayne Wise at *National Geographic Traveler,* who sent me on the initial "Meeting the Family" journey, and whose unerring sense of narrative resulted in the Lowell Thomas Award for us on the story. Also, Oliver Payne, Peter Miller, and Don Belt at *National Geographic,* plus Steve Perrine at *BestLife* (who exercised his usual gift for insight by sending me to the Chernobyl Exclusion Zone), not to mention the hilarious Alex Heard at *Outside;* the generous John Rasmus, Steve Byers, and Mark Adams at *National Geographic Adventure;* Bruce Handy and Graydon Carter at *Vanity Fair;* David Dibenedetto, Dave Mezz, Haskell Harris, and Sid Evans at *Garden & Gun;* and finally the wonderful Kathleen Burke, Beth Py-Lieberman, and Carey Winfrey at *Smithsonian.*

Around the homestead, thanks to a few friends and neighbors: Ed and Suzanne Chitwood, Ellie and Grice Whiteley, Donald and Anne McCaig, Doug and Sarah DuPont, Peter and Jane-Ashley Skinner, Jim and Leslie Bergin, Martin and Leslie Baruch, Garrick and Debbie Louis, and Matthew and Lori Blumberg.

Also thanks to the masterful Steve McCurry, the photographer who accompanied me on the original "Meeting the Family" trip to make the magazine illustrations, and whose photographs now grace the pages of this book. You're a great traveling companion, Steve. Let's find another story.

Beyond my children, several people and resources got me thinking about the larger implications of being human today, thoughts that eventually birthed this book. Chief among these is Michael Garstang, Ph.D., emeritus professor of environmental science at the University of Virginia and chairman of the National Academy of Sciences *Report on Weather and Weather Modification,* which got me mulling the scale of "deep time" and the scale and always shifting status of environmental systems on Earth, not to mention how life and volcanic and atmospheric changes affect these systems in return. Regards also to my friend David McCall in Canada, a careful and devoted reader who, once he heard what I was pondering, sent me a stream of books and thoughts on the subject.

Also valuable in shaping this book's Bigger Picture is the tone-perfect, poetically told history of war titled *Catapult: Harry and I Build a Siege Weapon* by Jim Paul (Villard, 1991), which I've been re-reading every year since its publication and—until recently—didn't know why. Still, it wasn't until I also read Bill Bryson's *A Short History of Nearly Everything* (Broadway, 2003) and Michael Pollan's *The Omnivore's Dilemma: A Natural History of Four Meals* (Penguin Press, 2006), that I was prompted to think: I'd like to try doing that with the human saga. And may my book be *half* as entertaining as both of yours.

Another profound Big Picture resource was Joseph Tainter's *A Collapse of Complex Societies* (Cambridge University Press, 1990), which embedded notions into this book's DNA, both optimistic and pessimistic, about human invention, creation, and chaos. Equally valuable was *A Short History of Progress* by Ronald Wright (Anansi, 2004), which provided navigation points, including the notion of many prehistoric societies not leaving "black boxes" so we could reconstruct their crashes, plus the idea of "Progress Traps," which I've massaged into the concept of technology traps, believing as I do that human advancement is not all inherently bad. The book *Catching Fire: How Cooking Made Us Human* by Richard Wrangham (Basic, 2009) was eye-opening in its notion that cooking is what separates us from the apes, and in its examination of the ways early humans made huge technological leaps that have now become so central to our lives they're almost invisible.

And, finally, in the Big Think jurisdiction, there are two more: *The Times Atlas of World History* (Jeffrey Barraclough, ed., Times Books, 1978), which has lived beneath my roof since I was a young man and still remains fantastic in its characterization of broad-brush human trends and migrations, and *The Upside of Down: Catastrophe, Creativity, and the Renewal of Civilization* by Thomas Homer-Dixon (Island Press, 2008), which lays out how human, energy-based, social, and environmental "tectonic stresses" accrue over time, shoving humans into corners they previously couldn't have imagined—sometimes leading to new forms of human advancement. That is, if we can figure out how to manage these transitions successfully.

For Chapter 1, "Africa," I need to thank Julius and all the Hadzabe, Linda Van Horn at Northwestern University's Feinberg School of Medicine, and Polly Wiessner at the University of Utah. Materials used in the research of that chapter include *The Journey of Man* by Spencer Wells (Random House, 2003), *Deep Ancestry: Inside the Genographic Project* by Spencer Wells (National Geographic, 2006), and *Africa: A Biography of the Continent* by John Reader (Random House, 1997).

Regarding Chapter 2, "World of Possibilities," it's important to thank my colleague Virginia Morell at *National Geographic,* who one evening engaged me in a discussion that walked the dividing line between religion and science (while Rick Ridgeway looked on silently and grinned). It's a conversation that's defined how I've seen the world ever since. Also Polly Wiessner (again), Alison Brooks at George Washington University, and the work of Nicholas J. Conard at the University of Tübingen in Germany.

Books and resources useful for Chapter 2 include *Encyclopedia of Volcanoes* (Academic Press, 2000) edited by Haraldur Sigurdsson, *The Neanderthals: Changing the Image of Mankind* by Erik Trinkaus and Pat Shipman (Alfred A. Knopf, 1993), *The Last Neanderthal: The Rise, Success, and Mysterious Extinction of Our Closest Human Relatives* by Ian Tattersall (Westview Press, 1999), and *Origins Reconsidered: In Search of What Makes Us Human* by Richard E. Leakey and Roger Lewin (Doubleday, 1992). Also used was "Neandertals: The Dawn of Humans" by Rick Gore in *National Geographic* (January 1996) and the newspaper story "Flute Music Wafted in Caves 35,000 Years Ago" by John Noble Wilford in the *New York Times* (June 25, 2009).

For Chapter 3, "Lebanon," I need to thank Maya Tabarra and everyone in Baalbek, especially Mohamed Wehbe and Khalid Abbass. I am also grateful to Rae Lesser Blumberg of the University of Virginia and the staff and curators at the Museum of Anatolian Civilization in Ankara, Turkey.

Useful books and resources not cited above include *Monte Verde: A Late Pleistocene Settlement in Chile* edited by Tom D. Dillehay (Smithsonian, 1989), plus *New Light on the Most Ancient East* (Norton, 1969) and *Man Makes Himself* (New American Library, 1953) by V. Gordon Childe, and, finally, *Collapse: How Societies Choose to Fail or Succeed* by Jared Diamond (Penguin, 2006). Also used was the article "Coastal Exploitation" by Torben C. Rick and Jon M. Erlandson in the journal *Science* (August 2009), and the story "Ancient Man Hurt Coasts, Paper Says" by Cornelia Dean in the *New York Times* (August 21, 2009).

For Chapter 4, "Uzbekistan," it's impossible not to thank Dilshod and everyone at the Hotel President. Books and sources not cited above that were used for that chapter include the article "Economic Development and Crime" by Sethard Fisher in *The American Journal of Economics and Sociology* (July 2006 online), *The Ascent of Money: A Financial History of the World* by Niall Ferguson (Penguin Press, 2008), *A Green History of the World* by Clive Ponting (Penguin, 2007), and *Nature's Place: Human Population and the Future of Biodiversity* by Richard P. Cincotta and Robert Engelman (Population Action International, 2000).

For Chapter 5, "Spain," I extend gratitude to the Carlton Hotel and Enrique Cardenas at the Café Boulevard, Zahi Hawass in Egypt, Bill Calvin in Seattle, plus Sergei Ivanchuk and Near East Travels in Kiev and Vladimir Verbitskiy and Maria Shylan inside the Chernobyl Exclusion Zone.

New resources in this chapter include *Genghis Khan and the Making of the Modern World* by Jack Weatherford (Crown, 2004), *Flushed: How the Plumber Saved Civilization* by W. Hodding Carter (Atria, 2007), my own story "Valley of the Mummies" in *National Geographic* (October 1999), and "Tuna Town in Japan Sees Falloff of Its Fish" by Martin Fackler in the *New York Times* (September 20, 2009).

My last shreds of thanks and gratitude go to this book's editor, Susan Tyler Hitchcock at National Geographic Books, who initially heard my idea one evening during a get-together over some memorably good soup, and who clung to it even when it had been dismissed by other editors as a result of my being "out in the sun too long." I hope, Susan, you're half as pleased with the result as I am. You're the best. Thanks also to National Geographic cartographers Carl Mehler and Matt Chwastyk for developing and producing the map, and to Daniel O'Toole, who fact-checked the final manuscript.